TAKING BETTER CARE OF MONKEYS AND APES

Refinement of Housing and Handling Practices for Caged Nonhuman Primates

by VIKTOR REINHARDT

Animal Welfare Institute

Animal Welfare Institute
P.O. Box 3650
Washington, DC 20027
www.awionline.org

Copyright © 2008 by the Animal Welfare Institute
Printed in the United States of America

ISBN 0-938414-96-8
LCCN 2008922969

Cover photo by Viktor Reinhardt
Design by Ava Rinehart
Copy editing by Cat Carroll and Cathy Liss

Mixed Sources
Product group from well-managed
forests, controlled sources and
recycled wood or fiber
FSC www.fsc.org Cert no. SW-COC-002504
© 1996 Forest Stewardship Council

Table of Contents

1. *Introduction* 1-2

2. *Definitions*
   2.1. Refinement 3
   2.2. Distress 3-6
   2.3. Well-Being 6

3. *Signs of Refinement* 7

4. *Distressing Conditions* 9-98
   4.1. Barren Cage 9-53
      4.1.1. Signs of Distress and Impaired Well-Being 10-14
      4.1.2. Refinement 14-53
         4.1.2.1. Companionship 14-34
            4.1.2.1.1. Previously Single-Housed Animals Can be Transferred to Social-Housing Arrangements Without Undue Risks 15-24
            4.1.2.1.2. Compatible Companionship Enhances Well-Being by Addressing the Need for Social Contact and Social Interaction 25-26
            4.1.2.1.3. Companionship Buffers Fear and Anxiety 26-28
            4.1.2.1.4. Companionship Buffers Physiological Distress 28-30
            4.1.2.1.5. Companionship Promotes Health 31
            4.1.2.1.6. Companionship Alleviates or Eliminates Behavioral Pathologies 32-34

4.1.2.2. Grooming Opportunities 35-36
4.1.2.3. Foraging Opportunities 36-46
    4.1.2.3.1. Food Puzzles 38-41
    4.1.2.3.2. Food Dispensers 42-43
    4.1.2.3.3. Food with or on Substrate 43-46
4.1.2.4. Access to the Vertical Dimension 47-52
4.1.2.5. Environmental Enrichment 52-53

4.2. Separation from the Companion 53-58
    4.2.1. Signs of Distress and Impaired Well-Being 53
    4.2.2. Alternatives to Partner Separation 53-58
        4.2.2.1. Post Operative Recovery 54
        4.2.2.2. Food Intake and Metabolic Studies 54-55
        4.2.2.3. Neurophysiological Studies 55-58

4.3. Social Conflicts 58-62
    4.3.1. Signs of Distress 58-59
    4.3.2. Refinement 60-62
        4.3.2.1. Breaking Visual Contact 61
        4.3.2.2. Access to the Vertical Dimension of the Enclosure 61-62
        4.3.2.3. Careful Reintroduction after Separation 62

4.4. Enforced Restraint 62-89
    4.4.1. Signs of Distress 63-64
    4.4.2. Refinement 64-89
        4.4.2.1. Training to Cooperate during Injection and Venipuncture 64-75
        4.4.2.2. Training to Cooperate during Sample Collection
              from Vascular Access Ports 75-76
        4.4.2.3. Training to Cooperate During Saliva Collection 76
        4.4.2.4. Training to Cooperate During Semen Collection 77-78
        4.4.2.5. Training to Cooperate During Blood Pressure Measurement 78-81
        4.4.2.6. Training to Cooperate During Oral Drug Administration 81
        4.4.2.7. Training to Cooperate During Topical Drug Application 82
        4.4.2.8. Pole-and-Collar-and-Chair Training 83-86
        4.4.2.9. Training to Cooperate for Weighing 86-87
        4.4.2.10. Training to Cooperate for Capture 88-89

5. *Discussion*

    5.1. Compatible Companionship 91-95

    5.2. Foraging Opportunities 95-96

    5.3. Access to the Vertical Dimension 96-98

    5.4. Positive Reinforcement Training 98-100

6. *Conclusions* 100-101

7. *References* 102-137

1. INTRODUCTION

Russell and Burch (1992) introduced the concept of the 3 Rs—Replacement, Reduction and Refinement—in their 1959 book, *Principles of Humane Experimental Technique*. The concept was endorsed by the biomedical research community in the 1980s, but only two of the 3 Rs—Replacement and Reduction—have received serious attention. The practical relevance of the third—Refinement—has largely been overlooked (Office of Laboratory Animal Welfare, 2002). A search of the literature shows that articles dealing with Replacement and Reduction by far outnumber those dealing with Refinement (**Figure 1**).

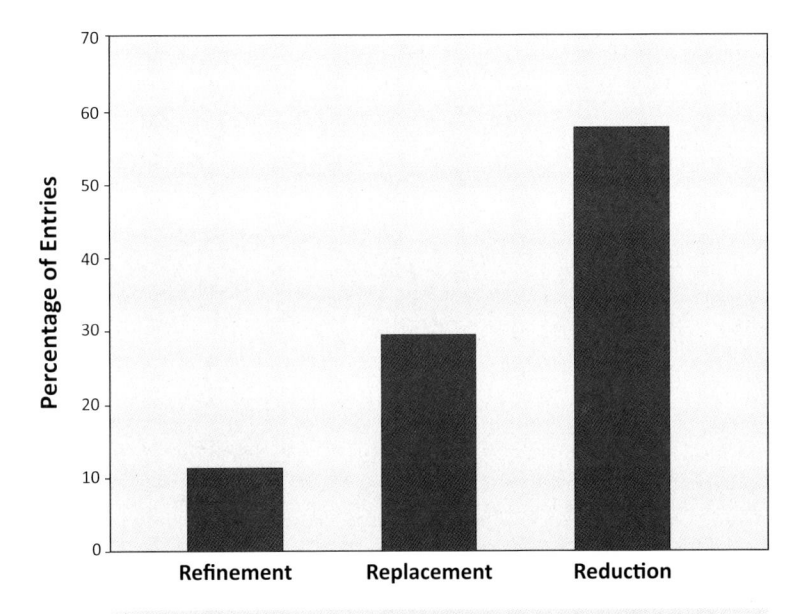

Figure 1. Results of a Scirus database search for the keyword string *Animal Testing Alternatives* & *Use of Laboratory Animals* & *Refinement/Replacement/Reduction* on June 30, 2007.

This book reviews the literature on the Refinement of traditional housing and handling practices for nonhuman primates who live in cages alone, in pairs or trios; articles dealing with group-housed animals (four or more animals) are not included. Published material has been reviewed if detailed data and sufficient information are provided that would allow the replication of the study in a different facility. Purely descriptive or theoretical material has not been included.

I am very grateful to my wife Annie Reinhardt, my daughter Catherine Reinhardt-Zacaïr, and the Animal Welfare Institute's Catherine Carroll and Cathy Liss for carefully checking the text and correcting grammatical errors and stylistic flaws.

It is my wish that the information compiled in this booklet will inspire animal caregivers, animal technicians, clinical veterinarians and researchers who are responsible for the welfare of caged primates to alleviate the animals' avoidable burden of distress.

Mt. Shasta, California   Viktor Reinhardt
January 2008

2. DEFINITIONS

2.1. Refinement

Russell and Burch (1992) define *Refinement* as:
> *Any decrease in the incidence or severity of inhumane procedures applied to animals* (p 65). *Its object is simply to reduce to an absolute minimum the amount of distress imposed* (p 134).

Balls et al. (1995), Buchanan-Smith et al. (2005) and Russell (2005) extended this definition by emphasizing that Refinement enhances the subject's *well-being*.

In the present review, the term "refinement" is used for:
> *Any modification in the housing and handling practices of animals that*
> - *reduces or eliminates the subject's distress response to a specific condition (e.g., permanent single-housing) or situation (e.g., enforced restraint during a life-threatening procedure), and/or*
> - *enhances the subject's well-being.*

2.2. Distress

In this review, *distress* is interpreted as:
> *Inability to adapt to a condition or to a situation that induces an alteration in the subject's physiological and psychological equilibrium.*

The following gestures and behaviors are taken as indicators that a nonhuman primate is distressed:
- Retreating to an upper back corner, crouching in the back of the cage, alarm vocalizing, fear-grinning, aggressive yawning, and self-biting in

Figure 2. Rhesus macaque *Betty* is quasi-cornered as personnel approach her cage. She responds with fear, anxiety and defensive aggression to this distressing situation. Note that *Betty* has lost part of her hair (alopecia) as a result of compulsive hair-pulling.

response to a potentially life-threatening situation (e.g., personnel approaching the cage). The subject is in a state of *anxiety* because a harmful event may happen, and frustration because there is no option to escape (**Figure 2**).
- Fear-grinning, struggling, and urinating in response to being forcefully restrained. The subject is in a state of *fear* because an uncomfortable or painful event is about to happen, and *frustration* because there is no option of escape (**Figure 3**).

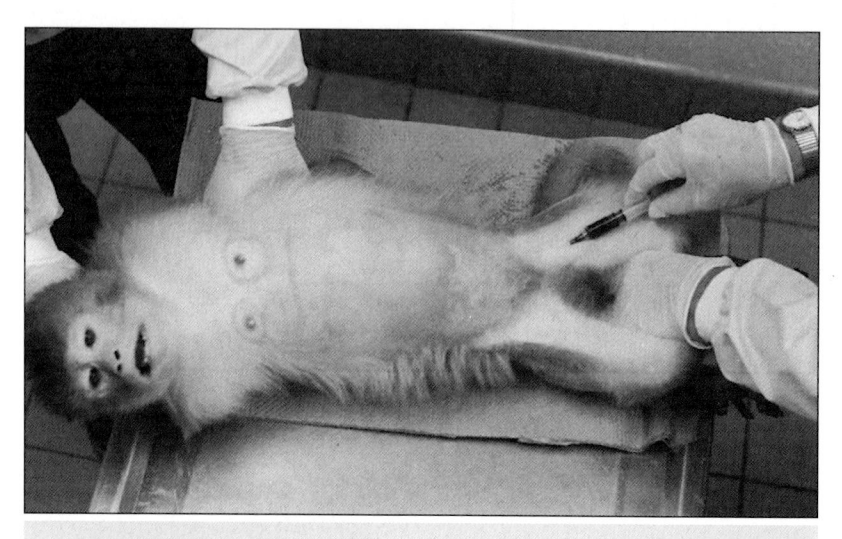

Figure 3. Rhesus macaque *Ella* is subjected to enforced manual restraint during routine blood collection. *Ella* exhibits signs of intense fear, indicating that she is distressed.

- Self-biting. This behavioral pathology occurs under the following circumstances:
  (a) Stereotypic self-biting

  The subject is extremely bored, shows no signs of excitation, and repeats the same movement patterns over and over again—for example, circling, pacing, bouncing or somersaulting—interjected by sham biting of specific body parts (**Figure 4a,b**). This behavior often goes unnoticed because there is no visible abrasion or laceration, and the subject usually does not show the behavior when there is a distraction—for example, when personnel is present.

  (b) Compulsive self-biting

  The subject is extremely frustrated—with high emotional arousal, e.g., shaking, intense staring, piloerection—for example, when fear-inducing personnel approach the cage, with the subject having no option to escape or attack. The animal will predictably bite specific parts of his or her body, such as always the

Figure 4a,b. This juvenile male rhesus macaque shows a behavioral distress reaction to permanent confinement in a barren cage. He bit his upper arms, wrists and thighs 636 times during a 60-minute video recording. Each "attack" lasted from a split second to as long as six seconds.

right wrist or always the left upper thigh. This leads to noticeable abrasion over time—first, local alopecia, followed by mild inflammation—but may also result in serious injuries. Typically, an animal self-inflicts lacerations of the same body part several times on different occasions (**Figure 29a,b**), often necessitating the amputation of the repeatedly injured limb.

Self-biting and other forms of self-injurious behaviors also occur in human primates in association with depression, anxiety and incarceration (Scott and Gendreau, 1969; Sluga and Grünberger, 1969; Wells, 1974; Bach-Rita, 1974; Yaroshevsky, 1975; Villalba and Harrington, 2003).

- Hair-pulling. The subject pulls single hairs or tufts of hair from his or her own fur or from the fur of a cage mate, manipulates the hair with the fingers, lips and tongue, chews the hair and finally ingests it. Hair-pulling often leads to localized alopecia (**Figure 2**).

  Hair-pulling is also relatively common in humans (Ko, 1999). It is classified as a *mental disorder* [*trichotillomania*] (Hallopeau, 1894), associated with *clinically significant distress* (American Psychiatric Association, 1987) *depression, frustration, boredom, or other emotional turmoil* (Christenson and Mansueto, 1999). It stands to reason that hair-pulling in nonhuman primates is also a sign of distress.

- Depression in response to being harassed by the cage mate. The subject consistently avoids the partner and spends most of the time crouching in a corner of the cage (**Figure 50**).

In this review, repetitive gestures (e.g., saluting), behaviors (e.g., ear-pulling) and movements (e.g., pacing) without obvious function [stereotypies] are not being considered as unequivocal indicators of distress, even though they reflect species-inadequate housing conditions.

2.3. Well-Being

In this book *well-being* is defined as:

A state of ease in which the subject's needs for survival are met.

For nonhuman primates in professionally accredited research facilities, the *physiological* needs are usually met while the *behavioral* needs for survival are often not addressed. This review, therefore, focuses on well-being that is derived from the performance of behaviors that would be crucial for the subject's survival in the wild.

3. SIGNS OF REFINEMENT

Refinement is successful if it:
- buffers distress as reflected in a reduction or elimination of self-biting or hair-pulling;
- buffers distress as reflected in the reduction of fear, anxiety and frustration;
- enhances well-being by providing species-adequate opportunities for the expression of behaviors that have a distinct survival value:
    a) being with and interacting with another conspecific (social behavior);
    b) searching for, retrieving and processing food (foraging); and
    c) accessing high refuge areas (vertical flight response).

Manipulating objects or toys, gnawing inedible objects, and looking into mirrors and monitors have a temporarily entertaining effect, rather than survival value. Since it is questionable that the performance of such behaviors enhances well-being, they have not been included as signs of refinement in this review.

Figure 5. Nonhuman primates such as baboons have a biologically inherent need to be in the company of conspecifics.

4. DISTRESSING CONDITIONS

4.1. Barren Cage

Solitary imprisonment is a dreaded punishment for human primates, who suffer from apathy, depression, frustration and behavioral pathologies when they are kept alone on a long-term basis (Scott and Gendreau, 1969; Sluga and Grünberger, 1969; Wells, 1974; Bach-Rita, 1974; Yaroshevsky, 1975; Walters et al., 1963; Grassian, 1983; Suedfeld, 1984; Grassian and Friedman, 1986; Gamman, 1995; Andersen et al., 2000; Andersen et al., 2003; Arrigo and Bullock, 2007). It stands to reason that nonhuman primates, who are also highly evolved social creatures, suffer when they are forced to live permanently alone in barren cages.

Figure 6a,b. Solitary imprisonment is distressing not only for human primates (a), but also for nonhuman primates (b).

Seeing the inside of a primate research facility for the first time was a shocking experience for me, not only as a psychologically healthy person but also as a scientist who has been trained to rigorously control extraneous variables that might influence research data. There were hundreds of animals kept in barren single-cages with nothing to do but stare at bleak walls and wait for their turn to be subjected to life-threatening procedures (Reinhardt and Reinhardt, 2001).

4.1.1. Signs of Distress and Impaired Well-Being

Being permanently imprisoned in a barren cage is distressing and impairs the well-being of nonhuman primates for the following reasons:

1. Primates have a biological need for companionship (**Figure 5**). Without other conspecifics, a monkey or ape has no chance of long-term survival in the wild. To be with and interact with at least one companion is a fundamental condition for the well-being of primates. When they are kept alone on a permanent basis primates tend to:

    (a) suffer from apathy, depression (**Figure 6a,b**; Luck and Keeble, 1967; Erwin and Deni, 1979; Lilly et al., 1999), extreme boredom and frustration (**Figure 7**) resulting in the development of compulsive hair-pulling and self-biting (**Figure 2 & 4a,b**; Erwin et al., 1973; Gluck and Sackett, 1974; Anderson and Chamove, 1981; Russell and Russell, 1985; Line et al., 1990; Watson, 1992; Platt et al., 1996; Lutz et al., 2000a; Kaufman et al., 2002; Marshall et al., 2002; Tully et al., 2002; Novak, 2003; Baumans et al., 2007), and

    (b) become more susceptible to disease (Shively et al., 1989; Reinhardt, 1990a; Schapiro

Figure 7. *Hatty* has been imprisoned in a barren cage for many years. The hyper-aggressive gesture suggests that *Hatty* is frustrated with her species-inappropriate living condition.

and Bushong, 1994; Poole et al., 1999).

2. In their natural habitat, nonhuman primates spend a major portion of the day foraging (**Figure 8**). They have a biologically inherent need to do so; it keeps them alive. Even though primates kept in research laboratories have no real need to forage, since their daily food ration is usually freely presented, they are strongly motivated to work for their food anyway. Experiments conducted with gibbons (Markowitz, 1979), stump-tailed macaques (Anderson and Chamove, 1984; Washburn and Rumbaugh, 1992; O'Connor and Reinhardt, 1994; Chamove, 2001), long-tailed macaques (Evans et al., 1989; Watson et al., 1999), rhesus macaques (Line et al., 1989; Reinhardt, 1994a), chimpanzees (Menzel, 1991), vervet monkeys (Pastorello, 1998) and marmosets (de Rosa et al., 2003; Bjone et al., 2006) have revealed that the animals will spend a considerable amount of time and effort to retrieve food that is hidden behind a barrier, even though the same food is also freely accessible next to them. From this, it can be inferred that they are highly motivated to forage, with the engagement in foraging activities serving as primary reinforcement.

Figure 8. In their natural habitat, baboons and all other nonhuman primates spend a major portion of their time foraging, i.e., searching for, retrieving and processing food.

Foraging has a distinct survival value for primates. Therefore, it can be assumed that the animals' well-being is promoted when they are given the opportunity to engage in food searching, processing and food retrieving activities.

3. In the wild, nonhuman primates spend the night and a major portion of the day well above the ground in trees, on rocky outcroppings or cliffs. Access to the vertical dimension is a basic condition for them to escape and to be safe from predators during periods of affiliative and playful social interaction, rest and sleep (**Figure 9a**). Most primates also forage in trees (**Figure 9b**). Without access to the vertical dimension, they are restricted to a terrestrial lifestyle to which they are not adapted (**Figure 10**).

Figure 9a,b. Nonhuman primates are arboreal animals; a) Vervet monkeys; b) Rhesus macaques.

Figure 10. Nonhuman primates are not adapted to a terrestrial lifestyle, yet these rhesus macaques are imprisoned in a bottom-row cage without elevated structure.

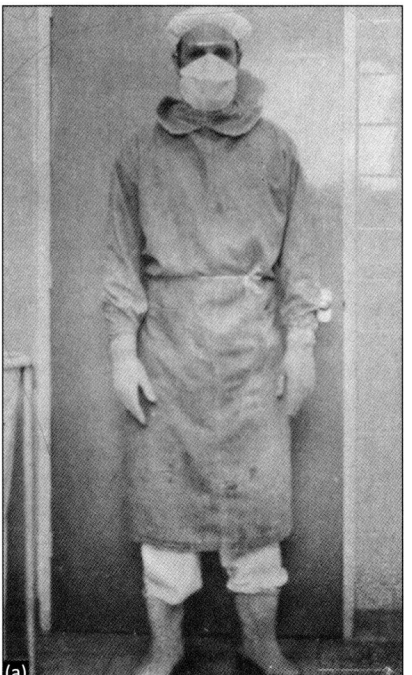

Figure 11a,b. The female rhesus macaque at right feels distressed because a fear-inducing investigator (a) is approaching her cage, and she has no option to retreat to a high, quasi-safe refuge (b).

Figure 12. In their natural habitat, macaques spend a major portion of their time grooming each other.

When they are confined in barren cages with no possibility of retreating to a high, safe place, nonhuman primates are literally cornered when they are approached by human primates who, after all, are their natural predators. This common situation is likely to distress the animals in research laboratories on a daily basis (**Figure 11a,b**).

4.1.2. Refinement

Refining aspects of housing, husbandry, enrichment and socialization helps alleviate or prevent distress (National Research Council, 2008, p 55).

4.1.2.1. Companionship

In the wild, primates benefit from each other's survival skills, such as avoiding predators, fleeing from predators, and finding species-appropriate foodstuff. A socially isolated primate would have no chance of long-term survival. Primates have a strong need for companionship. Taking the example of capuchin monkeys, it has been demonstrated that the animals perceive a companion as a *necessity* at a level similar to that of food (Dettmer and Fragaszy, 2000). Their social disposition is underscored by the observation that individually caged animals often try to touch and interact with their neighbors, despite substantial physical restriction and no visual access (Chamove, 1989, **Figure 7**; Baker, 1999).

Studies of wild populations indicate that Old World primates spend 5 to 25 percent of the day interacting with each other, with grooming being the prevalent social activity (**Figure 12**; *long-tailed macaques*: Leon et al., 1993; McNulty et al., 2004; *rhesus macaques*: Lindburg, 1971; Teas et al., 1980; Chopra et al., 1992; *Japanese macaques*: Hanya, 2004; *chimpanzees*: Wrangham, 1992; *baboons*: Hall and De Vore, 1965). Comparative data on New World primates have yet to be published.

4.1.2.1.1. Previously Single-Housed Animals Can be Transferred to Social-Housing Arrangements Without Undue Risks

Line et al. (1990) established four pairs of previously single-caged adult female **long-tailed macaques** (cynomolgus macaques, *Macaca fascicularis*) by introducing the potential companions in double-cages without any preliminaries. All four pairs were compatible and no fighting occurred during a two-week follow-up period.

Crockett et al. (1994) pre-familiarized the partners of 15 adult female and 15 adult male long-tailed macaque pairs via transparent cage dividers, allowing visual (but not physical) contact. After two weeks, pairs were formed by removing the divider. On the first day of introduction, partners were separated after 90 minutes. On each of the next 12 days, they were housed together for seven hours and separated during the remaining 17 hours to allow for collection of urine samples. Under these circumstances, only 53 percent of the male pairs turned out to be compatible. Within the first two weeks, 47 percent (7/15) of them had to be separated because of repeated fighting and serious lacerations. None of the female pairs had to be separated; they were all compatible.

Lynch (1998) applied a less disruptive pair formation strategy to 34 adult male long-tailed macaques. Potential partners were also first given the opportunity to get to know each other during a non-contact familiarization period, but they were introduced to each other—in a different double-cage to avoid possible territorial antagonism—only after they had established a dominance-subordinance relationship. Once paired, they were allowed to stay together uninterruptedly throughout the day and night. Under these conditions, 94 percent (16/17) of the pairs turned out to be compatible over follow-up periods of 12 to 42 months (**Figure 13**). Serious fighting at the time of introduction occurred in only one incompatible pair.

Clarke et al. (1995) established a trio of previously single-caged adult male long-tailed macaques by:
1. Exposing each subject to a mirror to provide an intermediate form of social stimulation during a two-week period.
2. Exposing each male to each other in a pair-wise arrangement that allowed visual, auditory and olfactory access to each other, but no opportunity for physical contact during a two-week period.
3. Introducing the three males into a group cage, one at a time, in rapid succession.

Figure 13. Long-tailed macaques *Ted* and *Tom* have lived together as compatible companions for more than three years.

The formation of the trio was not associated with serious fighting. Group members spent much of the time grooming each other during the first two weeks, and relationships between them appeared to be relaxed. The primarily affiliative and submissive behaviors shown by the three males suggest that they were able to establish a dominance hierarchy and harmonious relations quickly and easily. They were living peacefully together during a follow-up period of three years.

Byrum and St. Claire (1998) established 12 pairs of previously single-caged adult female **pig-tailed macaques** (*Macaca nemestrina*) after partners had established dominance-subordinance relationships during a one-week non-contact familiarization period. No injurious fighting occurred, neither at the moment of introduction nor during a two-year follow-up period.

Gust et al. (1996) released eight previously single-caged adult female pig-tailed macaques and one adult male simultaneously into a compound and encountered no problems. The animals established dominance-subordinance relationships within the first week without engaging in overt aggressive interactions.

Reinhardt et al. (1988a) placed previously single-caged adult female **rhesus macaques** (*Macaca mulatta*) pair-wise in double-cages, with partners being separated from each other by a wire mesh partition permitting non-contact communication. The animals were familiarized in this manner for seven days. Partners were paired on day eight only if they had not been seen threatening each other across the grated

Figure 14. Food sharing is one factor that distinguishes rhesus macaques *Sissi* and *Jill* as compatible companions five years after pair formation.

partition. The actual introduction then took place in a different double-cage to avoid the risk of possible territorial antagonism. A total of 27 dyads were tested. Partners threatened each other during the familiarization situation in nine (33 percent) of the cases. Reciprocal threatening was not witnessed in the other 18 dyads and the partners were, therefore, paired with each other. They were compatible in 83 percent (15/18) of cases during a follow-up period of five to six years. Absence of serious aggression, as well as food sharing distinguished partner compatibility (**Figure 14**); this implied that subordinate animals showed the same body weight gains, as did their dominant partners (Reinhardt et al., 1988b). Pairs were incompatible in 17 percent of cases, with one animal inflicting a serious injury on the other in one case, and one partner showing signs of social distress in the other two cases. These three dyads were permanently separated on days four, five and 15, respectively.

Subsequent work with female and male rhesus showed that the two partners of compatible pairs do not differ in their serum cortisol concentrations, indicating that living with a compatible companion does not constitute a distressing situation for either the subordinate or the dominant partner (**Figure 15**; Reinhardt et al., 1990a; Reinhardt et al., 1990b). The same findings have been made in squirrel monkeys (Gonzalez et al., 1982), and they may apply to all other primate species when animals are housed on a long-term basis as compatible pairs.

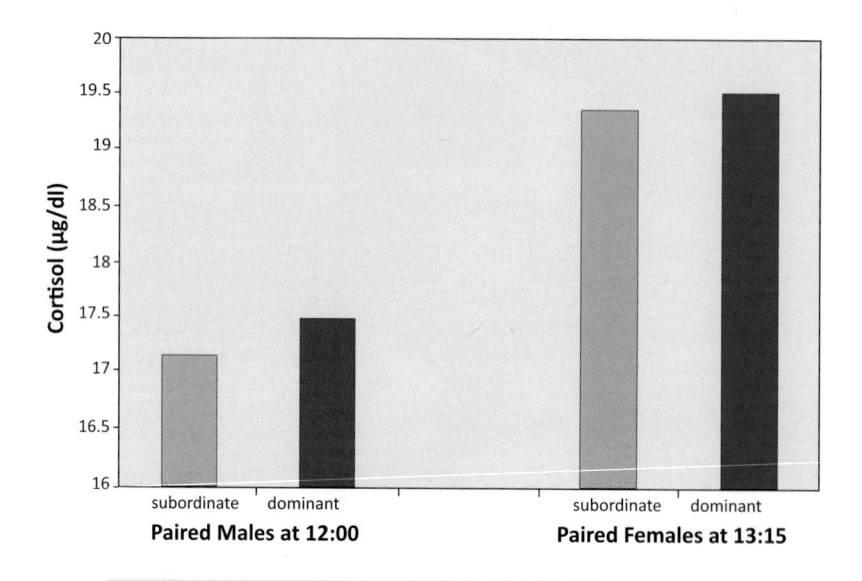

Figure 15. Mean serum cortisol concentrations of the dominant and subordinate partner of five compatible male and five compatible female adult rhesus macaque pairs. The animals were trained to cooperate during venipuncture; blood samples were taken from the males at 12:00 and from the females at 13:15 (Reinhardt et al., 1990b).

Eaton et al. (1994) applied a similar pre-familiarization technique with female rhesus. Of 21 pairs tested, 86 percent (18/21) were compatible throughout a follow-up period of more than three years, and 14 percent of the pairs were incompatible and had to be separated because of serious fighting during the first hour (two cases) or after three months (one case). The partners of compatible pairs spent 40 percent of the time during the day in close proximity, and 80 percent of the time during the night. They did not show any differences in body weight gains, clinical morbidity, reproduction and immune response. This suggests that subordinate animals were not hindered by their dominant companions to obtain the appropriate share of the daily food ration, nor was their health and general well-being jeopardized by their dominant cage mates.

In order to minimize the potential risk of injurious fighting, Reinhardt (1989a) refined this pair formation protocol for adult rhesus males by making it a condition that potential partners must establish a dominance-subordinance relationship during non-contact familiarization, so that they will have no reason to fight over dominance when they are introduced to each other. Seven pairs were tested. Two of them failed to establish a clear-cut dominance-subordinance relationship. Five did establish such a relationship, with one of the partners showing unidirectional submissive gestures. When the partners of these five pairs were introduced to each other in a different double-cage, not a single incident of fighting occurred, and the animals reconfirmed their already established rank positions with subtle gestures involving no physical contact (**Figure 16**).

This pair formation technique was subsequently implemented at a research facility as a standard procedure for adult rhesus macaques, including 24 to 35 year old animals (Reinhardt, 1991b; **Figure 17**). When 77 female pairs and 20 male pairs were established on this occasion, fighting occurred in only 2 percent of the 97 pairs: two female pairs and no male pair (Reinhardt, 1994b).

Doyle et al. (2008) familiarized the potential partners of four adult rhesus macaque pairs in cages in which partners were separated by a panel consisting of bars spaced 2 cm apart. The eight males were all implanted with biotelemetry devices for remote heart rate monitoring. After 24 hours, as neither persistent aggression nor wounding was observed, each pre-familiarized pair was introduced into full contact by removing the barred panel. All four introductions were successful and subjects showed no physiological (fecal cortisol concentration and heart rate) or behavioral signs (pathological behavior) of stress, or psychological indices of distress (depressive/anxiety-related behavior) not only during the introduction process but also over a follow-up period of 18 months. No overt aggression was displayed at all during the first two hours following pair formation. Aggressive interactions were minimal thereafter. Only one bite laceration was incurred 14 weeks after pair formation. The partners of this pair were maintained in the home cage with the barred panel to allow wound healing; they were subsequently placed

Figure 16. Rhesus macaque *Mike* grooms his dominant cage mate *Bob* after they have reconfirmed their rank relationship with subtle gestures.

again into full contact with no further complications.

Roberts and Platt (2005) paired adult rhesus males who had cranial implants. Potential companions were familiarized and their compatibility was carefully evaluated over a period of five weeks. In order to be physically introduced in the same test cage, partners had to establish a clear-cut dominance-subordinance relationship during the first week, when the animals were separated by transparent cage dividers. During the next four weeks, partners were allowed to live together intermittently for progressively longer periods of time. After the fifth week, they finally lived together continuously. Of 13 pairs tested in this manner, 92 percent (12/13) were compatible. Only one pair was deemed incompatible because of continued non-injurious aggression during the sixth week. This pair was separated.

Figure 17. Twenty-six-year-old *Sissa* grooms her 35-year-old companion *Senila* shortly after pair formation. These two aged rhesus macaques have lived most of their lives alone in barren single-cages.

Reinhardt et al. (1987) and Reinhardt (1991) examined the practicability of pairing adult rhesus macaques with infants. Naturally weaned, 12 to 18 month old infants of both sexes were removed from two breeding troops to avoid overcrowding and placed, without any preliminary precautions, pair-wise with unfamiliar single-caged, 7 to 33 year old adults of both sexes. A total of 40 pairs were tested: 12 adult female-infant female pairs, 11 adult female-infant male pairs, 11 adult male-infant male pairs, and six adult male-infant female pairs. The pairs were compatible in 92 percent (37/40) of cases with:
- the adult protectively holding the infant **(Figure 18a,b)**,
- the infant showing no signs of depression **(Figure 18a,b)**
- the infant being able to get his or her share from a limited amount of favored food **(Figure 19a-d)**, and
- the adult inflicting no visible injury on the infant.

Compatibility was dependant neither on the sex of the adult and infant, nor on the age of the adult partner. Three pairs were incompatible. One female grabbed the female infant immediately upon her arrival; she continued to do this repeatedly during the next 30 minutes, after which the infant was removed. One male bit the female infant on the fourth day of introduction. The youngster was slightly injured, although not bleeding. When the infant started to consistently avoid the adult, the pair was split. Another male often grabbed his male infant companion, even though he gently groomed him and the two huddled with each other regularly. Gradually, however, the infant showed more and more avoidance behavior, and the two were finally separated after nine days.

Several attempts have been described to transfer single-caged adult rhesus macaques to compatible group-housing arrangements, but none of them were successful enough to be recommended as a safe standard procedure. Whether future group members are strangers or have been carefully pre-familiarized with each other, and whether they are introduced simultaneously or sequentially as a new group, vicious and even deadly fighting and persistent aggressive harassment seem to be unavoidable (Bernstein and Mason, 1963; Erwin, 1979; Jensen, 1980; Line et al., 1990a; Reinhardt, 1991b; Clark and Blanchard, 1994).

Figure 18a,b. Rhesus macaques *Matt* (a) and *George* (b) hold and huddle their infant cage mates *Jimmy* and *Billy*, who show no signs of depression. Both males are very protective of their little companions; they yawn because they feel uncomfortable being observed.

Figure 19a-d. Adult rhesus macaque *Cora* allows her infant companion *Gina* to get her share of food treats. Note that *Gina* has a cranial implant.

Reinhardt (1994c) transferred 10 adult female and six adult male **stump-tailed macaques** (*Macaca arctoides*) from single-housing to isosexual pair-housing by first allowing potential partners to establish dominance-subordinance relationships without risk of injury during a three-day non-contact familiarization phase, and then introducing them to each other in a new home cage. All five female and all three male pairs established clear-cut dominance relationships while they were familiarized with each other. Following subsequent introduction, all eight pairs showed signs of compatibility. Female partners reconfirmed their rank relationships within 30 minutes with subtle gestures, never by overt aggression. Male partners engaged in *hold-bottom rituals* (de Waal and Ren, 1988) upon being introduced to each other. The partners of two pairs reconfirmed their rank relationships within 30

Figure 20. Stump-tailed macaques *Roger* and *Paul* get along well with each other six months after pair formation.

minutes with gestures, while the third pair resorted to a brief non-injurious dominance reconfirming fight which was followed by another reconciliatory hold-bottom ritual. The eight pairs remained compatible, with no signs of injurious aggression throughout a six-month follow-up period (**Figure 20**).

Bourgeois and Brent (2005) established four pairs and two trios of previously single-caged 3 to 4 year old male **baboons** (*Papio* sp.) by sedating potential companions and having them wake up together in the same cage. *Rough-and-tumble wrestling* occurred and dominance positions were quickly established, with all disputes followed by bouts of grooming. Transfer to social-housing was successful in each instance, and no injuries or overt aggression were observed during a follow-up period of two weeks (**Figure 21**).

Figure 21. These three baboons are a compatible trio.

Fritz and Fritz (1979) and Fritz (1994) developed a protocol to introduce previously single-caged **chimpanzees** (*Pan troglodytes*) to unfamiliar peers. The newcomer is first moved into a specially designed *social unit* and kept next to the cage of a selected member of an already established group. The two chimpanzees have full olfactory, visual and auditory contact as well as limited tactile contact. The selected group member is moved in as a cage mate for the newcomer as soon as friendly interactions through the separating cage mesh are consistently observed. After several days, another group member is introduced to the pair in this same way, then another is introduced to the trio, and so on until the newcomer has met all members of the group and is then fully integrated. A total of 59 of 60 chimpanzees—of both sexes and all age classes—were successfully re-socialized to compatible group-living in this manner without a single incidence of serious fighting (**Figure 22**; Fritz, 1989).

Gwinn (1996) used a pole-housing system to identify compatible adult male **squirrel monkeys** (*Saimiri* sp.) before introducing them as pairs:

> Pole-housing allows several primates to interact or retreat to safety. First the animals are habituated to collar, leash and pole. During this time, the animals cannot physically interact with others. When they have adapted to the pole system, they are moved closer to one another. They are observed for aggression or fighting at frequent time intervals. When two animals exhibit compatibility, having been observed interacting positively for one week, they are pair housed. Eight monkeys are currently housed as pairs.

The percentage of pairs exhibiting compatibility in the pole-housing arrangement is not indicated.

Figure 22. Living in a compatible group allows previously single-caged chimpanzees to express their social needs.

4.1.2.1.2. Compatible Companionship Enhances Well-Being by Addressing the Need for Social Contact and Social Interaction

Compared to wild animals, captive pair-housed primates spend more time engaged in social activities—especially grooming each other—probably because there is little else for them to do.

Reinhardt and Reinhardt (1991) kept 15 adult female **rhesus macaque** pairs in double-cages that were each equipped with a *privacy panel* allowing the partners to stay in different halves of the cage without maintaining visual contact with each other. During one-hour observations, companions spent 76 percent of the time in the same half of the cage. Obviously, they had a need for companionship and preferred not to be alone, even though this implied a relative reduction of the available cage space. They were engaged in grooming and hugging each other on average 37 percent of the time.

Basile et al. (2007) observed 25 adult female rhesus pairs in double-cages with privacy panels for two 30-minute sessions. Companions spent 52 percent of the time in the same half of the cage, and they engaged in affiliative interactions 24 percent of the time.

Eaton et al. (1994) established 11 pairs of adult female rhesus macaques and recorded their behavior during 10-minute sessions, three times per week during a six-month period. Companions spent on average 35 percent of the time engaged in species-typical social behavior, with grooming being the predominant interaction (31 percent). There was no indication that companions lost interest in each other over time.

Ranheim and Reinhardt (1989) took two 30-minute behavioral records of six pairs of adult female rhesus macaques who had lived together for 30 months. Companions spent on average 35 percent of the test sessions interacting with each other, primarily in the form of grooming (30 percent). Apparently, partners had not become bored with each other during the two and a half years of uninterruptedly living together in the same cage.

Reinhardt and Hurwitz (1993) paired three 30 to 35 year old female rhesus macaques—who had lived most of their lives alone—with compatible adult female partners. During three one-hour sessions conducted 16 months after pair formation, the three aged animals were grooming and hugging their companions on average 29 percent of the time (**Figure 17**).

Baker (2007) observed 13 adult male rhesus pairs during 12 half-hour sessions. Partners spent an average of 18 percent of the time in affiliative interactions.

Line et al. (1990a) formed five pairs of adult female **long-tailed macaques.** During approximately seven hours of observation distributed over the first two weeks, partners spent approximately 31 percent of the time grooming each other.

Crockett et al. (1994) recorded the behavior of 15 female and 8 male pairs of adult long-tailed macaques 13 days after the pairs were formed. During a 90-minute test session, female companions spent an average of 35 percent of the time while male companions spent an average of 17 percent of the time grooming each other (**Figure 13**).

Reinhardt (1994c) established five pairs of adult female and three pairs of adult male **stump-tailed macaques.** During one-hour observations conducted six months later, females interacted with each other on average 24 percent of the time, males interacted with each other 17 percent of the time (**Figure 23a,b**). Grooming (77 percent) and hugging (22 percent) were the salient social activities.

4.1.2.1.3. Companionship Buffers Fear and Anxiety

Like human primates (Arsenian, 1943; Schachter, 1959; Wrightsman, 1960), nonhuman primates have a reassuring, anxiety-reducing effect on each other in distressing situations.

Rowell and Hinde (1963) exposed 17 rhesus macaques of both sexes and all age classes to a *mildly stressful* situation, i.e., being looked at by a person with a grotesque mask, for three minutes alone or with several familiar group members. When they were tested alone, the animals showed significantly more signs of fear (threatening, hair raising), anxiety (yawning) and tension (scratching) than when they were exposed to the stressor in the company of other monkeys.

Gunnar et al. (1980) captured five infant rhesus macaques from their social group and placed them in an unfamiliar environment for 24 hours, either alone or with another infant from the same group. When tested alone, the animals exhibited significantly more signs of distress (agitation and distress vocalization) than when they were tested with a companion, indicating that the companion had a stress-buffering effect.

Mason (1960) placed 12 infant rhesus macaques into a strange environment,

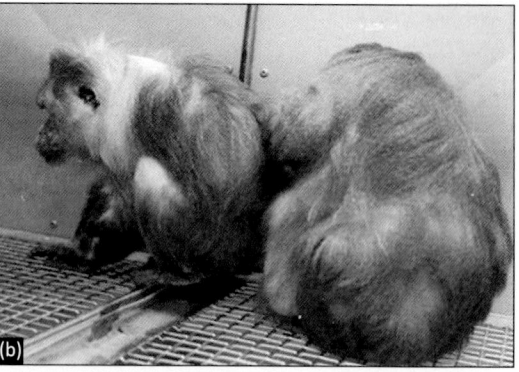

Figure 23a,b. Stump-tailed macaques *Claudia* and *Clara* are engrossed in reciprocal grooming.

either alone or with another familiar or unfamiliar same-aged peer. Subjects showed significantly fewer signs of emotional disturbance (crouching and self-clasping) when they were tested in the company of another monkey. The distress-buffering effect was not dependent on the familiarity of the accompanying partner.

Due to repeated traumatic experiences with humans, caged monkeys often become alarmed when a person enters the room (Malinow et al., 1974; Manuck et al., 1983; Hassler et al., 1989; Arluke and Sanders, 1996; Capitanio et al., 1996; Schnell, 1997; Bowers et al., 1998; Boinski et al., 1999; Crockett and Gough, 2002; Lueders, 2004). During such frightening situations, paired animals often exhibit behavioral responses that suggest that they reassure and calm one another (**Figures 24a-c**).

Hennessy (1984) observed eight pair-housed squirrel monkey infants when they were transferred to an unfamiliar cage alone or with the companion. The animals vocalized significantly less when they were tested together, suggesting that the companion moderated the fear response to the unfamiliar environment.

Coe et al. (1982) confronted 14 adult squirrel monkeys for 60 minutes with a snake behind a mesh barrier and noticed that the animals'

Figure 24a-c. Rhesus macaques *Bobby* and *Circle* comfort each other while an investigator catches another animal in the room for an experimental procedure.

behavioral distress responses (alarm vocalization, fear reactions and agitation) were significantly reduced when they were tested in company of another male than when they were tested alone.

4.1.2.1.4. Companionship Buffers Physiological Distress

The physiological stress and distress response to challenging situations is mitigated by a social partner in human primates (Kissel, 1965; Epley, 1974; Lynch et al., 1977; Witcher and Fisher, 1979; Drescher et al., 1980; Kamarck et al., 1990; Gerin et al., 1992; Lepore et al., 1993; Gerin et al., 1995; Kirschbaum et al., 1995; Uchino et al., 1996; Christenfeld et al., 1997; Thorsteinsson et al., 1998; Fontana et al., 1999; Gallo et al., 2000; Uno et al., 2002). This seems to be true also for nonhuman primates.

Vogt et al. (1981) confronted 24 adult squirrel monkeys, who lived in four heterosexual groups, with a caged snake alone versus in the company of the other group members. The adrenocortical activation evoked by such a potent fear stimulus was significant when the animals were tested alone, but it did not occur when they were tested as a group.

Gonzalez et al. (1982) exposed six single-housed and six pair-housed adult female squirrel monkeys to the stress of capture followed by anesthesia and cardiac puncture, and found that the 30-minute plasma cortisol increment was significantly lower in subjects housed with a companion (38 percent) than in subjects housed alone (60 percent).

Coelho et al. (1991) measured blood pressure via arterial catheter implants of four tethered adult male baboons who were kept in a test room either alone or in a double-cage in which they had visual, tactile and auditory contact with a familiar companion through a wire mesh partition. Mean resting blood pressures were consistently lower when the baboons were able to interact with a neighboring baboon, suggesting that companionship buffered distress arising from imprisonment in an unfamiliar environment (**Figure 25**).

Doyle et al. (2008) assessed fecal cortisol levels and monitored heart rates of eight adult biotelemetry device-implanted male rhesus macaques (a) after they had lived alone in single-cages for several months and (b) after they were paired with each other and had lived together for more than four months. Both stress/distress parameters were significantly lower in the pair-housing versus the single-housing condition, indicating that the males experienced less distress in the company of another male than when they lived alone.

Gust et al. (1994) transferred seven adult female rhesus monkeys from their group to an unfamiliar environment, either alone or together with a preferred group member. During both conditions, subjects were initially equally distressed, as measured in

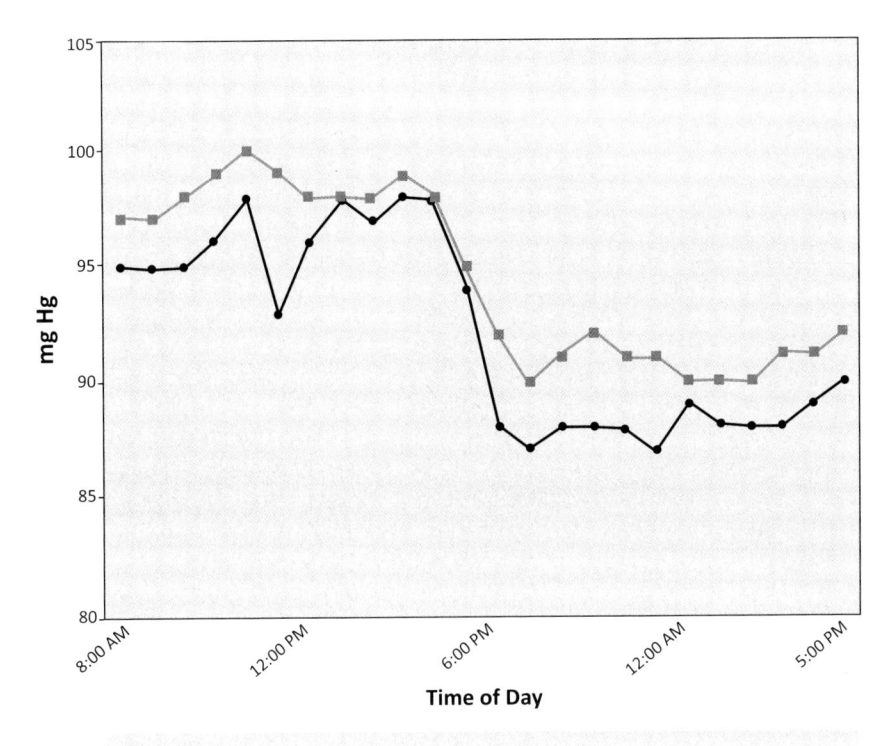

Figure 25. Mean arterial blood pressures of four tethered baboons when caged alone (gray line) versus with social contact (black line) in an unfamiliar environment (Coelho et al., 1991).

alterations of cell-mediated immune parameters, but they recovered significantly quicker when they had the social support of a companion.

Drug testing can be a distressing experience that is often reflected in the subjects' gradual loss in body weight. Gwinn (1996) noticed during nine treatments with an identical test compound that adult male squirrel monkeys lost significantly less weight when they were caged with a companion (n=4) than when they were caged alone (n=4).

It has been demonstrated in some species, especially human primates, that contact with friendly individuals of another species can have a calming, stress- and distress-buffering effect (Gantt et al., 1966; Lynch and Gantt, 1968; Lynch et al., 1974; Astrup et al., 1979; Hemsworth et al., 1981; Friedmann et al., 1983; Baun et al., 1984; Wilson, 1987; Vormbrock and Grossberg, 1988; Siegel, 1990; Allen et al., 1991; Barnett et al., 1994; Pedersen et al., 1998; Allen et al., 2001; Allen et al., 2002; Barker et al., 2005; Coppola et al., 2006; Cole et al., 2007) and enhance resistance to pathophysiological processes (Friedmann et al., 1980; Nerem et al., 1980; Todd-Schuelke et al., 1991/92; Anderson et al., 1992; Friedmann and Thomas, 1995; Craig et al., 2000; Cole et al., 2007).

There seems to be a general consensus that positive contact—not necessarily tactile contact—with personnel has a stress-mitigating effect on nonhuman primates in research laboratories (**Figure 26**; Anchel, 1976; Wolfle, 1987; Institute for Laboratory Animal Research, 1992; Canadian Council on Animal Care, 1993; National Research Council, 1998; American Association for Laboratory Animal Science, 2001; Bayne, 2002; Prescott, 2002; Primate Research Institute, 2003; Abney et al., 2006; Baumans et al., 2007). Studies have yet to be published to provide supportive data for this very plausible assumption.

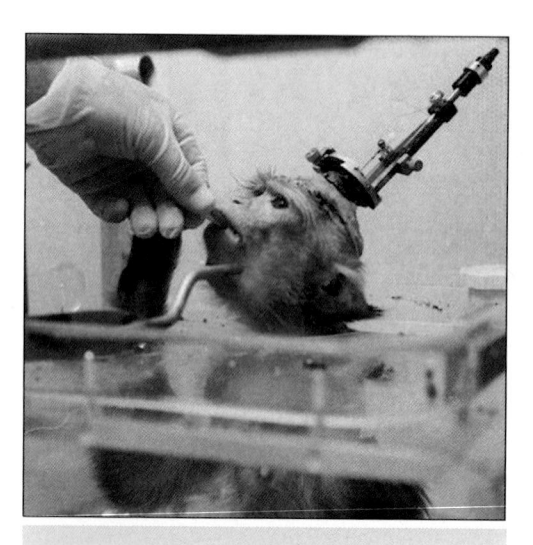

Figure 26. Regular affectionate interaction with attending personnel fosters a trust-based human-animal relationship that is likely to help the animal subject cope with distressing situations, such as being chair-restrained during a neurophysiological experiment.

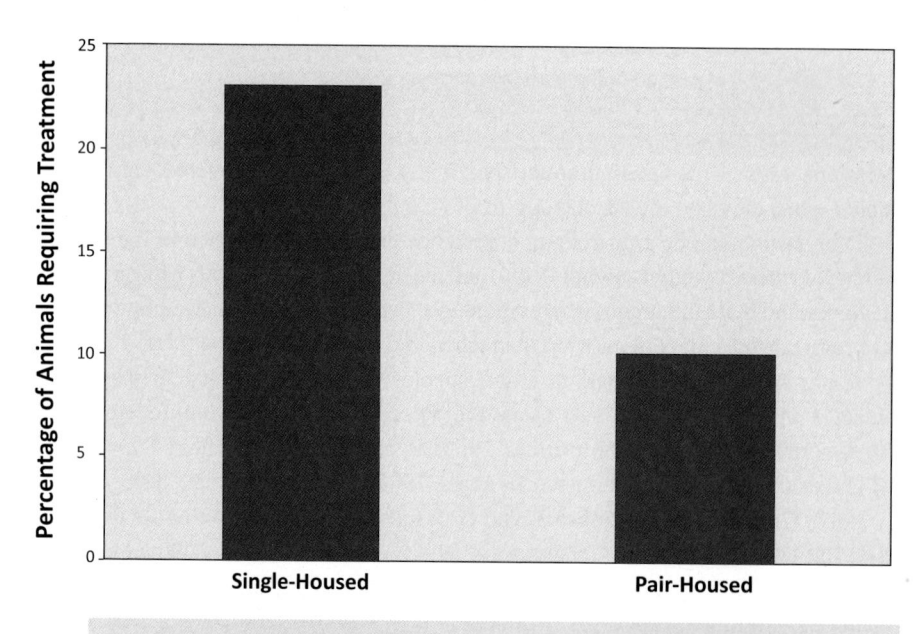

Figure 27. Percentages of a colony of 237 single-housed and 382 pair-housed rhesus macaques requiring veterinary treatment in the year 1989 (Reinhardt, 1990a).

4.1.2.1.5. Companionship Promotes Health

Schapiro and Bushong (1994) examined the health records of 98 rhesus macaques who were 1 to 2 years old when they were individually caged; they were 2 to 3 years old when they were subsequently kept in opposite-sex pairs; they were 3 to 4 years old when they were finally kept as breeding groups or male-only groups. Veterinary treatments were necessary:
- 39 times when the animals were caged alone,
- 17 times when they lived with a companion, and
- 55 times when they lived in groups.

The incidence of veterinary treatment was conspicuously low when the animals were pair-housed. This was probably related to the fact that *pair housed monkeys required significantly fewer veterinary interventions for diarrhea than did single or group housed monkeys* (Schapiro et al., 1997, p 147), and fight injuries requiring treatment were relatively common when the animals lived in groups. In a subsequent study, Schapiro et al. (2000) compared the cell-mediated immune response of 12 adult rhesus macaques who lived either alone, in pairs, or in breeding groups. Based on significant differences in the animals' immunological responses, it was contended *that strong social relationships, particularly the affiliative interactions that characterize pair housed monkeys, may diminish the likelihood of severe infection with potentially diarrhea-inducing agents* (p 79).

Reinhardt (1990) assessed the clinical records of a rhesus macaque colony consisting of 237 single-housed and 382 pair-housed animals of both sexes and all age classes. The incidence of non-research-related veterinary treatment was more than twice as high for single-caged than for pair-housed animals (**Figure 27**), indicating that the animals were healthier when they lived with a companion.

Shively et al. (1989) compared clinical data of female long-tailed macaques consuming an atherogenic diet and housed either alone (n=15) or with three or four other females (n=24). The extent of atherosclerosis was four times greater, on average, in females who lived alone than in those living with social companions (**Figure 28**). These findings corroborate with human primate studies demonstrating that lack of *social support* (House et al. 1982) is associated with an increased risk of coronary heart disease (Manuck et al., 1986; Lynch, 1987; Shumaker and Czajkowski, 1994) and other health issues (Kaplan et al. 1977; Berkman, 1985; Cohen and Syme, 1985; Broadhead et al., 1983; House et al., 1988; Christenfeld and Gerin, 2000; Hays et al., 2001; Spiegel and Sephton, 2001; Richmond et al., 2007).

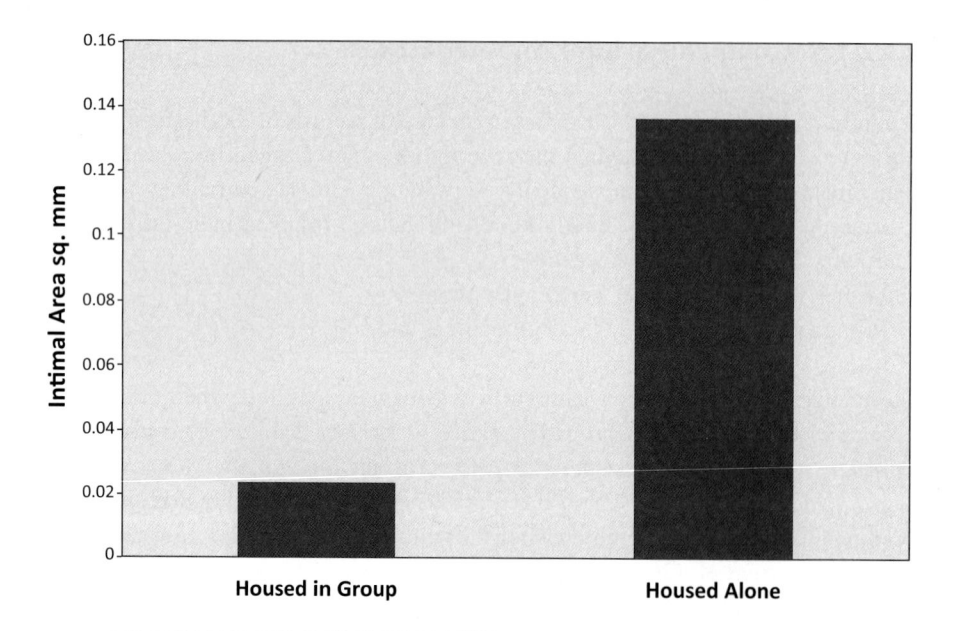

Figure 28. Mean coronary artery atherosclerosis extent as measured by intimal area in group-housed and single-housed adult female long-tailed macaques (Shiverly et al., 1989).

4.1.2.1.6. Companionship Alleviates or Eliminates Behavioral Pathologies

In a colony of about 650 mother-reared, single-caged adult rhesus macaques, **self-biting** was witnessed in four males and three females. This behavior pattern was predictably exhibited whenever one of the subjects was approached by personnel; the animal would show signs of intense excitation and start repeatedly biting a particular body part while staring and/or charging at the person. The self-biting resulted in no visible trauma in one female and two males; two females showed abrasions on the bitten hand; two males required surgical treatment, one of a lacerated thigh, the other of a lacerated arm. All seven subjects were successfully transferred from single- to compatible pair-housing arrangements with same-sex adult partners (six cases) or with an infant (one case). This had a therapeutic effect in all seven subjects: The conspicuous excitation and self-biting in the presence of personnel was abandoned immediately on the day of pair formation by three animals, or gradually within two months by the other four animals (**Figure 29a-c**). This pathological behavior pattern was no longer witnessed in any of the seven subjects (Reinhardt, 1999).

Weed et al. (2003) vasectomized six single-caged rhesus males, who engaged in persistent self-injurious biting, and paired them with adult females. Three of these males stopped the self-biting after being transferred to social-housing, and self-biting was no longer noticed during a one to six-month follow-up period. Socialization had a moderating but not healing effect in the other three males.

Alexander and Fontenot (2003) established 19 isosexual groups with 80 previously single-caged adult male rhesus macaques. Thirty-one (39 percent) of these animals had at least one prior incidence of self-injurious biting. During the year before group formation, the clinical history of the subjects included a 13 percent incidence of self-biting requiring wound care. No self-biting was noted during the first four months after the groups were formed.

Line et al. (1990a) paired five long-tailed macaques, who had a history of self-biting, with compatible female companions. Pair-housing corrected the behavioral problem and no further self-biting occurred in the course of a five-month follow-up period.

Reinhardt et al. (1987) transferred an adult female rhesus macaque from single-housing to pair-housing with a surplus infant from a breeding troop. While she was caged alone, *Chewy* predictably chewed and bit her left thumb whenever she was

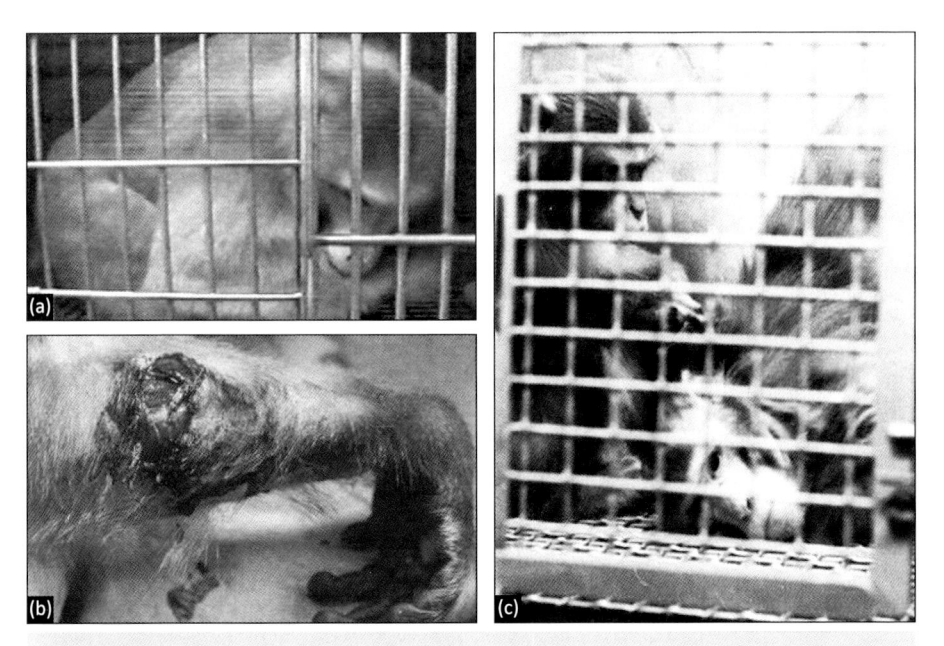

Figure 29a-c. Rhesus macaque *Paul* required two surgeries on self-inflicted bite lacerations (a,b). Being paired with *Peter* cured *Paul* of this behavioral pathology (c). In the course of a three-year follow-up period, *Paul* has not engaged in any noticeable self-biting.

approached by personnel. She stopped this compulsive behavior within the first month of living with her companion *Cute* (**Figure 49b**), and she did not resume it during a one-year follow-up period.

Baumans et al. (2007) refer to a case of three self-biting adult male rhesus:
The animals were treated with various drugs—diazepam, fluoxetine, guanfacine—which did alleviate but not eradicate the self-biting. Once the treatments were discontinued, the animals resorted to self-injurious biting (SIB) as before. All three males self inflicted repeatedly serious lacerations that required surgical care. When it was considered to euthanize these males, because the SIB could not be stopped with pharmacological therapy, permission was finally given to pair them with other compatible companions. This "treatment" brought the self-biting to an end in all three cases. Carl, however, had a relapse when his buddy was removed for research-assignment reasons after 14 months. Fortunately, the investigator was considerate enough to drop the companion from the research protocol. Once re-united with his companion, Carl promptly stopped again biting himself.

Fritz (1989) reports of three male and one female individually housed **chimpanzees** who stereotypically mutilated themselves. The animals were carefully socialized in compatible group settings that caused all four of them to gradually stop injuring themselves.

Minkel (2007) gives an account of a long-tailed macaque who was cured from compulsive **hair-pulling** by being paired with another conspecific:

At a previous institution we had a cyno—"Grandpa"—who suffered from severe hair-pulling. He had removed practically all hair from his body; all that was left was a patch in the middle of his back that he could not reach! He was not shy about hiding his idiosyncratic behavior and would contort into strange positions to do it. The veterinarians tried various treatments to alleviate the problem to no avail. We tried all enrichment devices we could find; they would only keep him occupied for a day or so. We increased the space of his cage; no luck. We were reluctant to pair him as he was an older male who had been singly housed for so long, but there was no other treatment option left.

We tried two unsuccessful pairings and finally settled on a newly acquired juvenile male who was very rowdy and active; Grandpa was quite the opposite, relaxed and sedate. The little guy himself was on his second pair attempt; during his first one—all he did was try to start a fight. To our great relief the new pair worked out just fine. This truly "odd couple" got along great from the start. Grandpa responded correctly, brought the little guy in line, and actually perked up. The most surprising part, however, was that Grandpa stopped hair-pulling. He stopped completely, and all his hair had grown back in the course of several months.

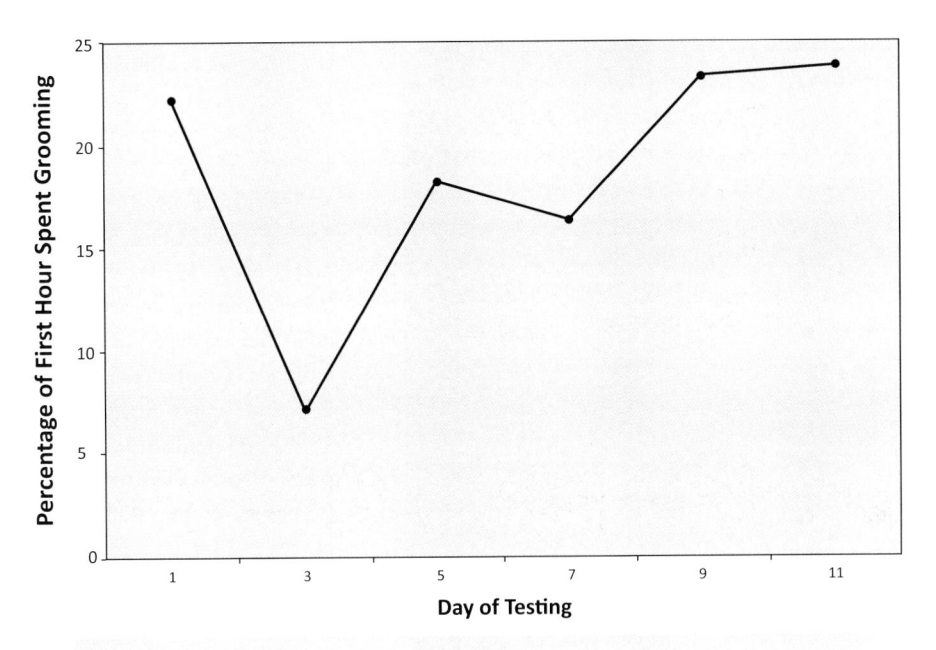

Figure 30. Mean grooming activity of three single-caged male long-tailed macaques who have 60-minute access to a grooming cushion every other day (Lam et al., 1991).

4.1.2.2. Grooming Opportunities

As an alternative to a social partner, Lam et al. (1991) gave three adult male long-tailed macaques a *grooming cushion*, consisting of a 20 x 20 x 60 cm large piece of synthetic fleece, every other day. The males would typically squat on the cage floor or sit on the perch and gently pluck at, stroke, or part small pieces of fleece with their fingers, just as they would do when grooming another monkey. This behavior was often accompanied by lip smacking. During one-hour observations, the animals spent on average 11 percent of the time grooming the cushion; there was no indication that they got tired of doing so in the course of an 11-day test period (**Figure 30**). A grooming cushion would probably provide suitable enrichment also for other primate species when individuals have to be caged alone for research- or health-related reasons.

Crockett et al. (1997) housed same-sex pairs of adult long-tailed macaques in double-cage units in which partners were separated by a blind panel for 19 hours daily. During the remaining five hours of the 24-hour day, they were separated by *grooming-contact bars*, allowing them to reach through with their arms. Of 16 female pairs tested, 100 percent were compatible and partners spent about 43 percent of the time grooming each other. Of 45 male pairs tested, 89 percent were compatible and partners spent about 7 percent of the time grooming each other (**Figure 31**).

Figure 31. Grooming-contact bars restrict paired companions to separate sections of the cage, but allow them to engage in species-typical grooming behavior. Here two adult male long-tailed macaques (*Macaca fascicularis*) in grooming-contact cages at the Washington National Primate Research Center.

Carolyn M. Crockett, Washington National Primate Research Center

The usefulness of grooming-contact bars or woven wire panels with mesh openings, large enough so that adjacent neighbors can groom each other (Coelho and Carey, 1990), has also been confirmed in adult iso- and heterosexual pairs of baboons (Coelho et al., 1991; Crockett and Heffernan, 1998) and adult heterosexual pairs of pig-tailed macaques (Crockett et al., 2001; Lee et al., 2005). Compared with other species, rhesus macaques do not adjust well to the grooming-contact housing system; paired animals show a relatively low incidence of compatibility, i.e., 16 percent versus 51 percent in pig-tailed macaques, 67 percent in long-tailed macaques and 64 percent in baboons (Crockett et al., 2006).

4.1.2.3. Foraging Opportunities

Perhaps the easiest way to allow primates to engage in food processing behavior is the daily provision of whole fruits, whole nuts and whole vegetables of the season—such as apples, bananas, oranges, grapes, ears of corn, celery, melons, pumpkins, sugar cane, etc. (**Figure 32a,b**). The common practice of chopping these supplemental food items deprives the animals of an opportunity to engage in a very important natural behavior. There are no published reports suggesting that the regular feeding of certain whole fruits, whole nuts or whole vegetables has any adverse side effects.

Numerous gadgets have been described to promote foraging in caged primates, but their actual effectiveness in promoting foraging—which does **not** include eating, i.e., ingesting food—for an extended period of time has been evaluated in only a few cases.

Figure 32a,b. Offering caged nonhuman primates whole fruits (a) and vegetables (b) allows them to engage in species-typical food processing activities.

Standard Feeder

Forage Feeder

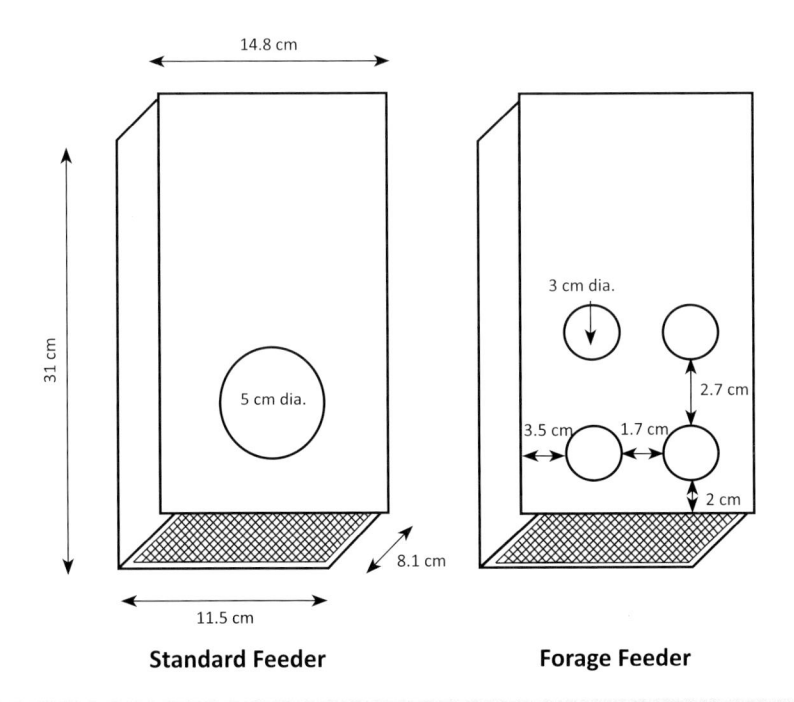

Figure 33. Reducing the size of the access hole (right) of the standard feeder (left) is a simple option for promoting skillful food retrieval behavior in nonhuman primates. Reproduced with permission from Murchison, M.A. 1995. Forage feeder box for single animal cages. Laboratory Primate Newsletter 34(1), 1-2.

4.1.2.3.1. Food Puzzles

Murchison (1995) designed a *forage feeder* for macaques that replaced the freely accessible standard feeder as primary source of the animals' daily biscuit ration. Both feeders were of the same dimension, but the puzzle had four holes, each 3 cm in diameter, while the box had a 5 cm-diameter opening, through which the much smaller biscuits could be directly picked up by the animals (**Figure 33**). During the first hour after distribution of biscuits in the food box or in the food puzzle, 20 adult female pig-tailed macaques spent on average: *1 percent (51 seconds) of the time collecting 44 biscuits from the box, versus 11 percent (400 seconds) of the time retrieving 44 biscuits from the puzzle.*

Reinhardt (1993a) re-mounted the two ordinary food boxes of eight pair-housed male rhesus macaques away from the 7.3 x 4.7 cm large opening right onto the 2.2 x 2.2-cm mesh of the front of the cages (**Figure 34a,b**). Skillful manipulations with the fingers were now required to maneuver each of the 4.0 x 2.4 x 1.6 cm large biscuits into the right position, break protruding parts off with the teeth or fingers and finally push-pull a biscuit through the mesh. The eight males received their daily ration of 66 biscuits in the early morning. Each pair was observed once when the ration was distributed in the ordinary food boxes, or in the two *food puzzles* to which the animals had first been habituated for 30 days.

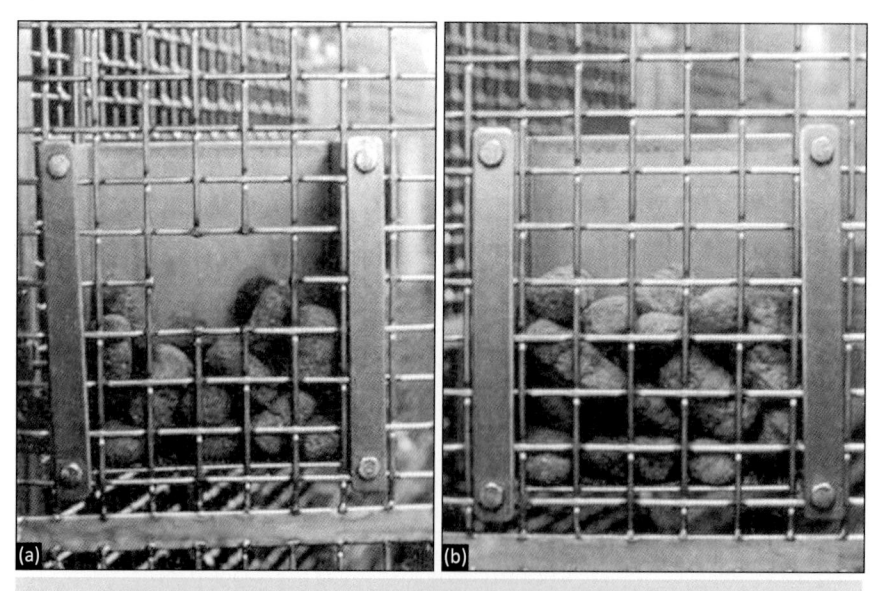

Figure 34a,b. Moving the ordinary food box away from the access hole (a) onto the mesh panel of the cage (b) will make it more difficult for the monkey to retrieve the food.

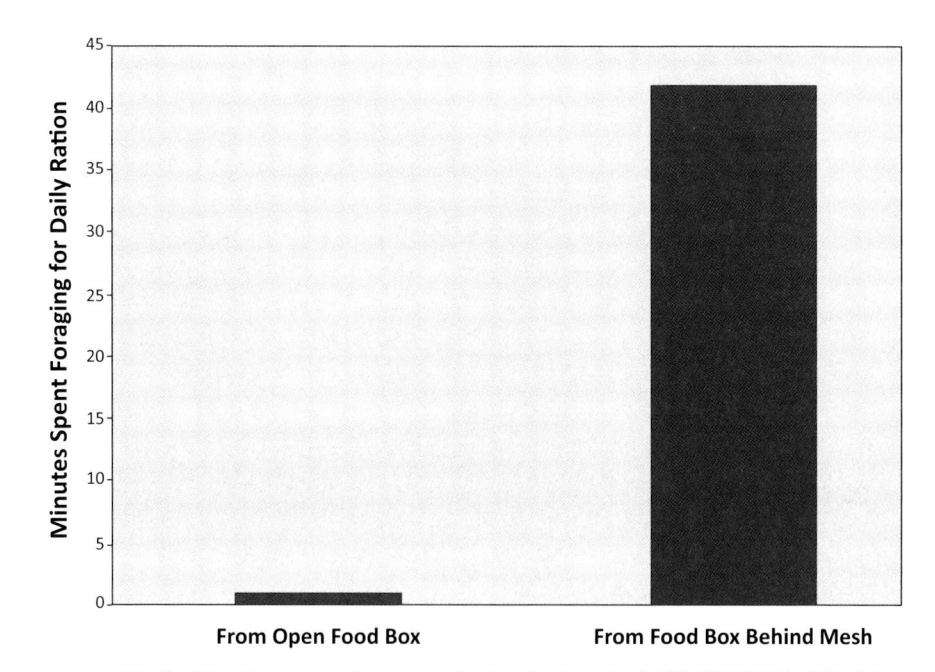

[Bar chart: Y-axis "Minutes Spent Foraging for Daily Ration" (0–45). "From Open Food Box" ≈ 1 minute; "From Food Box Behind Mesh" ≈ 42 minutes.]

Figure 35. Average time eight pair-housed adult male rhesus macaques spend foraging when their daily biscuit ration is placed in the ordinary food box mounted over the access hole versus directly onto the wire mesh panel of the cage (Reinhardt, 1993a).

When the standard biscuit ration was placed in the food puzzles instead of the food boxes:
- The average percentage of time spent foraging during the first 30 minutes increased significantly.
- The average total time spent collecting/retrieving 33 biscuits per animal increased significantly (**Figure 35**).

Working for the retrieval of their daily biscuit ration had no adverse effect on the males' body weight (Reinhardt, 1993b).

Reinhardt (1993c) tested this simple puzzle under the same methodological conditions in five adult single-caged female and seven adult single-caged male stump-tailed macaques (**Figure 36a-c**). When the 33-biscuit standard ration was placed in the puzzles instead of the food boxes:
- The average percentage of time spent foraging during the first 30 minutes increased significantly:
  (a) in females from 1 to 63 percent, and
  (b) in males from 1 to 62 percent.

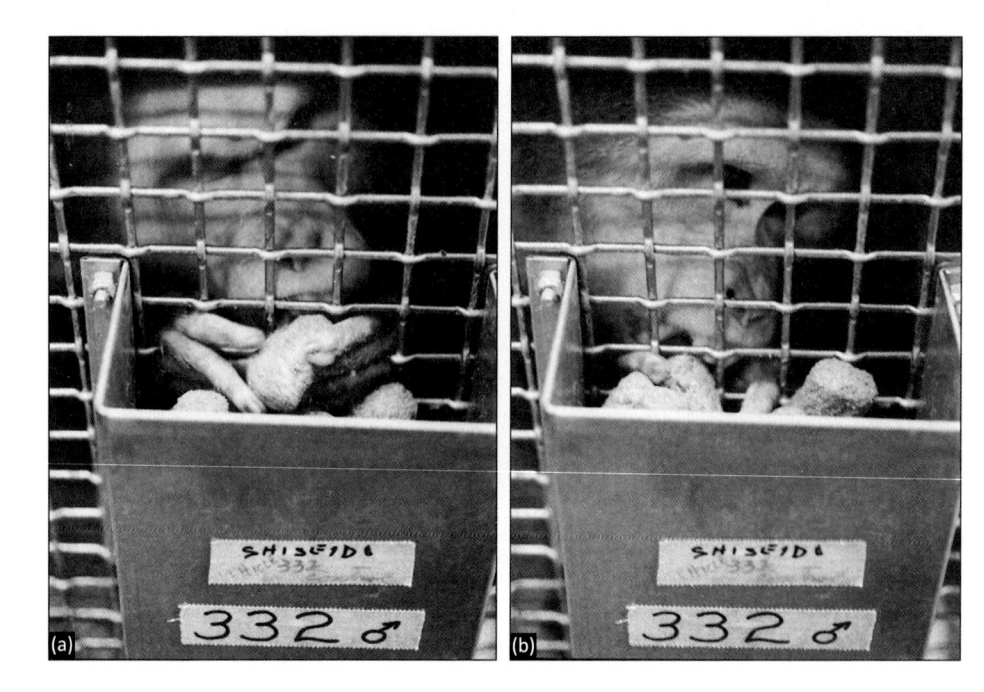

- The average time spent collecting/retrieving all 33 biscuits increased significantly:
  (a) in females from <1 to 31 minutes, and
  (b) in males from <1 to 23 minutes. The males retrieved the biscuits more quickly than the females, probably because they have stronger fingers and, therefore, can break biscuits and push them through the mesh more easily.

Foraging from the puzzle rather than collecting their daily biscuit ration from the food box did not affect the body weight development of the animals.

Glick-Bauer (1997) distributed the standard diet of an adult male cotton-top tamarin pair in the morning and again in the afternoon, either in an ordinary food dish, or in a 20 x 13 x 11 cm large plastic box with a hinged lid containing six 4 cm-diameter holes through which the subjects had to reach for and retrieve food items. During the first hour after food distribution, the two males spent on average:
- 4 percent of the time collecting food from the dish, versus
- 42 percent and 33 percent of the time retrieving food from the *puzzle feeder*.

Reinhardt (1993d) distributed the daily food ration, consisting either of 66 *small* bar-shaped or 32 *large* star-shaped biscuits, of eight pair-housed adult male rhesus macaques in their two ordinary food boxes, or on the 22 x 22 mm square mesh ceiling of the cage. The males had been habituated to both feeding options for a 12-day period. In the food box-situation, they had nothing to do but pick up one biscuit after the other; there was no effort involved. In the *ceiling puzzle* situation, the males had to maneuver each biscuit into the right position so that a part of it was protruding through the mesh, nibble or bite a piece off until the rest of the biscuit could be pushed with the fingers

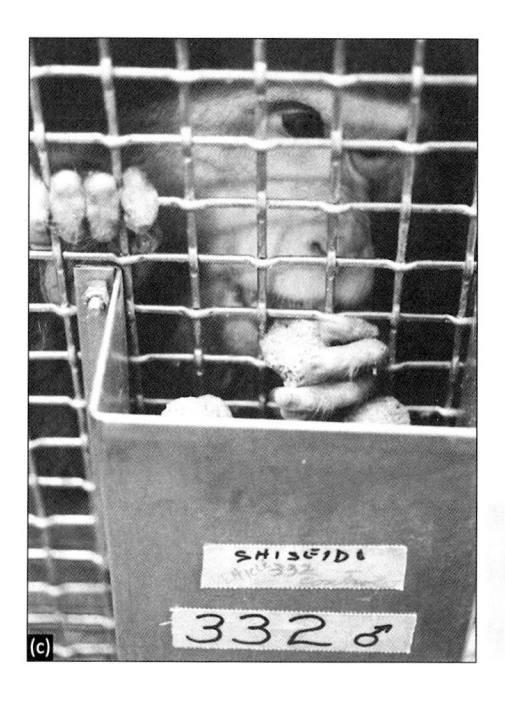

Figure 36a-c. Stump-tailed macaque *Steve* retrieves a biscuit of his daily ration through the wire mesh panel of the cage.

or pulled with the teeth through the mesh (**Figure 37a-c**). During the first four hours after biscuit distribution, the males spent on average:
- 0.3 minutes collecting all small and 0.2 minutes collecting all large biscuits from the food box, versus
- 23.0 minutes retrieving all small biscuits and 59.2 minutes retrieving most of the large biscuits through the mesh ceiling.

When the biscuits were presented in the open food box, the monkeys quickly took a few in their cheek pouches and threw many of the remaining ones onto the floor of the cage while starting to eat. When the biscuits were placed on the mesh ceiling, the animals ate all the retrieved pieces directly; they never stored them in the cheek pouches or threw them on the floor.

Bertrand et al. (1999) placed the daily biscuit rations of 12 individually housed rhesus macaques—of unspecified age and gender—for a period of two weeks, either in the ordinary freely accessible food box, in a container/puzzle mounted behind the mesh wall of the ceiling, or behind the mesh wall of the front of the cage. The two puzzles required skillful manipulations to retrieve the biscuits through the mesh. It took the animals on average about:
- 15 minutes to collect their ration from the food box, versus
- 60 minutes to retrieve their ration from the puzzle mounted on the front, and
- 75 minutes to retrieve their ration from the puzzle mounted on the ceiling of the cage.

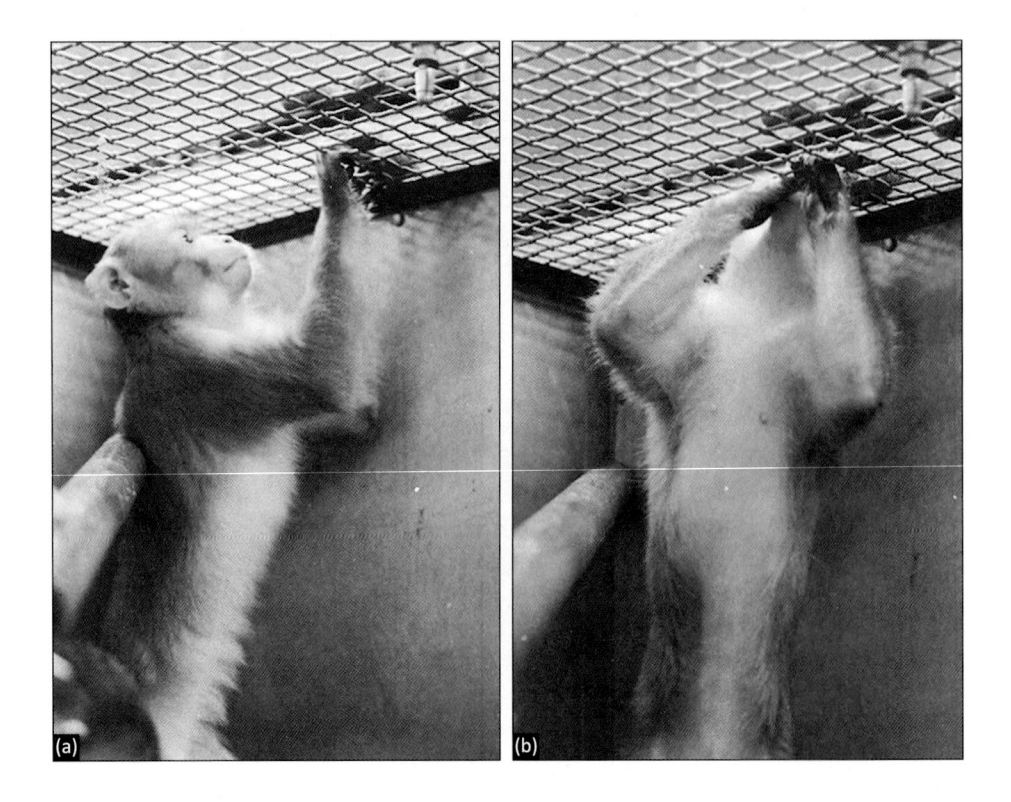

4.1.2.3.2. Food Dispensers

Bjone et al. (2006) exposed four adult female marmoset pairs twice daily for 20 minutes to a custom-designed feeder filled with standard food. The gadget was designed in such a way that the animals had to swing small discs to the left or right to uncover and retrieve food by reaching through little holes. The marmosets had simultaneous access to their ordinary open food bowls filled with the same food ad libitum. When given a choice between easily accessible food in a bowl and food from the puzzle, the marmosets predominantly chose to retrieve food from the puzzle. During six test sessions, they spent on average approximately:
- 0.2 minutes (1 percent of the time) collecting food from the bowl, versus
- 7.2 minutes (36 percent of the time) *foraging* at the puzzle.

Celli et al. (2003) mounted an open transparent polyethylene bottle, which was daily filled with honey, in front of the cage of three adult chimpanzee female pairs and offered the animals various materials, such as plastic brushes, wires, chopsticks and rubber tubes from which they could chose those suitable for retrieving honey from the bottle, similar to *fishing for termites* (Goodall, 1964) from termite mounds. During daily one-hour observations (probably right after presentation of the bottle) the animals spent on average:

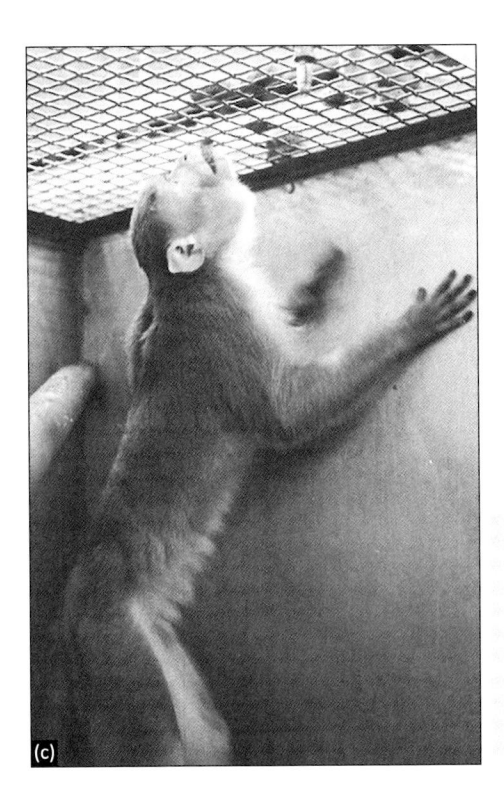

Figure 37a-c. Distributing the daily biscuit ration on the wire mesh ceiling of the cage, rather than in the standard open food boxes, allows macaques to engage in skillful foraging behavior. This kind of feeding enrichment is effective and does not cost anything.

- 9 percent of the time checking out suitable fishing tools, and
- 31 percent of the time *fishing for honey* (Paquette, 1992).

The chimpanzees engaged in these foraging activities consistently over the 10-day study period.

4.1.2.3.3. Food with or on Substrate

Bryant et al. (1988) released six individually caged adult male long-tailed macaques, one animal at a time, into a playpen on four consecutive days each week for a three-week study period and recorded their behavior 30 minutes prior to and 30 minutes after transfer to the pen. The monkeys were then returned to their home cages, where they received their normal food ration. The playpen was almost four times larger than the home cages and was furnished with a nylon ball, a telephone directory and a nylon rope, plus a tray placed below the grid floor of the cage, containing woodchips scattered with sunflower seeds and peanuts. The animals showed little interest in the enrichment items, but spent on average 33 percent (10 minutes) of the 30-minute observations in contact with the foraging tray, searching for and retrieving seeds and peanuts by

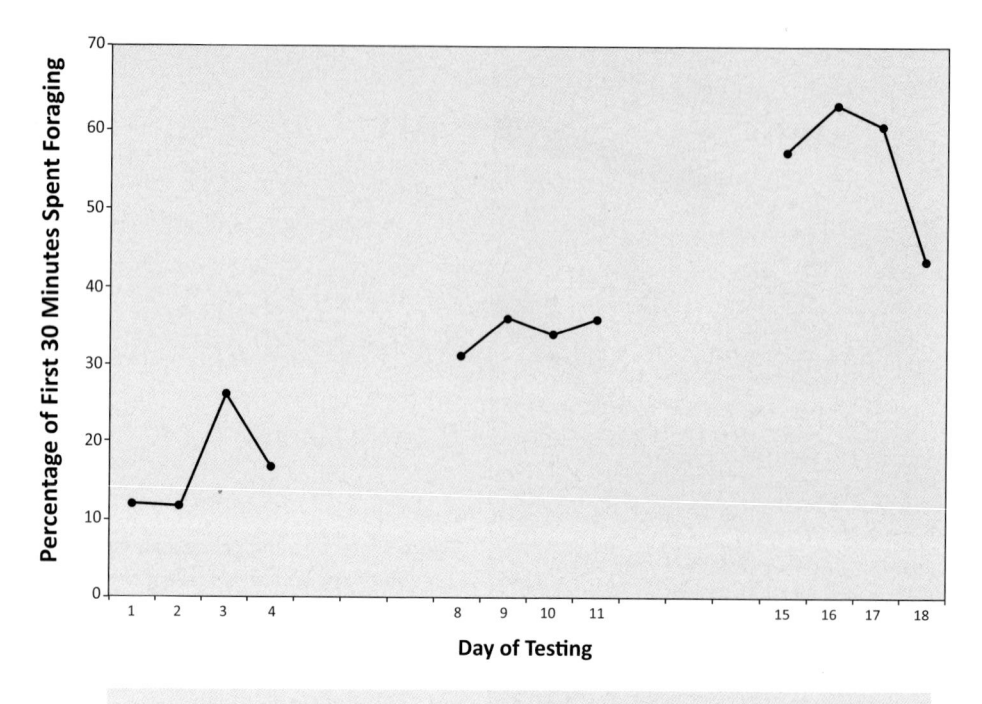

Figure 38. Mean foraging activity of six single-caged adult male long-tailed macaques who have daily access to a foraging tray placed beneath the mesh floor of the cage (Bryant et al., 1988).

reaching through the wire mesh of the cage floor. They increased their engagement in foraging in the course of the three-week study (**Figure 38**).

Baumans et al. (2007) quote an animal technician who distributes wood shavings sprinkled with sunflower seeds in the catch pans of rhesus and squirrel monkeys:

Our rhesus and squirrel monkeys search with their fingers through the litter and pull the seeds through the floor grids, eat them or store them in their cheek pouches. Since we change the pans three times a week, rather than dump the bedding, we don't have any drainage problems in the rooms. This feeding enrichment technique doesn't require undue extra work time in our colony of approximately 130 monkeys. I'd say the benefit of being able to provide even a brief period of "natural" foraging behavior for our caged primates is worth the little additional time it takes to put the bedding in the pans and add a handful of seeds.

Spector et al. (1994) furnished the drop pans of 24 single-caged baboons of unspecified age and gender with 29 x 44 x 6 cm large foraging trays. Every other afternoon, a mixture of seeds, dried fruits, pieces of vegetables, alfalfa cubes, feed corn and dog biscuits was added to the tray and then covered with a thin layer of fresh hay. The baboons had to reach through the bars of the cage floor, search for food items and then retrieve them. The animals were not systematically observed, but a review of many hours of video recordings taken during two years indicates that the baboons spent 30 to 120 minutes per day foraging from these trays.

Lam et al. (1991) gave three single-caged adult male long-tailed macaques each a 20 x 20 x 60 cm large synthetic fleece cushion sprinkled with favored tidbits before the regular feeding time on alternate days. The animals would sit on the perch or squat on the cage floor, picking out food crumbles with their fingers or directly licking the fleece. During the first 60 minutes after fleece cushion distribution, the males spent on average 40 percent (24 minutes) of the time foraging in this manner. They did not lose interest in foraging from the cushion over the course of a 12-day study period.

Bayne et al. (1992a) designed a 36 x 79 cm large foraging board consisting of Plexiglas covered with artificial turf. The board was secured to the cage floor and occupied approximately one third of it. Particles of flavored food items were sprinkled daily on the turf between the regular morning and afternoon feeding. These small tidbits sift down through the 13 cm long blades of the turf, thereby inducing an animal to engage in skillful manipulations to obtain the food (**Figure 39**). The board was tested in eight adult single-caged male rhesus macaques. It was replenished with food particles each day, after which the animals were observed for 30 minutes. During 20 sessions distributed over six months, subjects were occupied with foraging for an average of 52 percent of the time. Over the course of the study, the males increased the amount of time spent foraging from the turf board (**Figure 40**).

Lutz and Farrow (1996) secured 30 x 24 cm large turf boards to the outside of the front panel of the cages of ten adult female long-tailed

Figure 39. Female rhesus macaque *Boo* picks up tidbits from her foraging board that is attached outside to the front, rather than inside to the floor, of the cage.

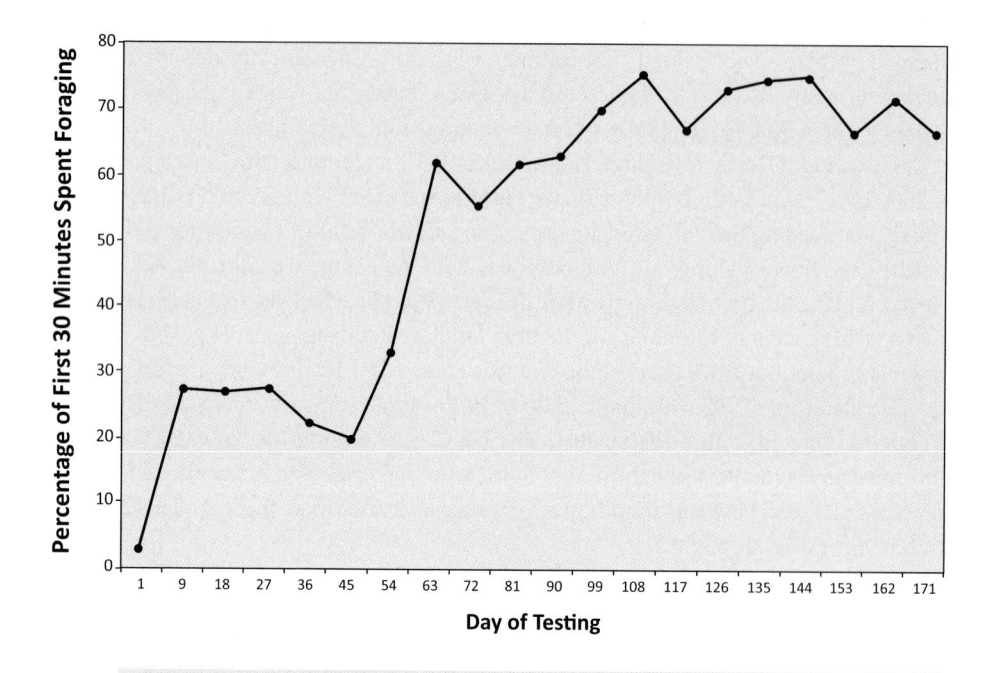

Figure 40. Mean foraging activity of eight single-caged rhesus macaques who have access to a turf board replenished daily with food particles (Bayne et al., 1992).

macaques and sprinkled sunflower seeds on the turf every morning after the animals had received their daily biscuit ration. During three weekly 30-minute observations conducted at random times over a period of eight weeks, the animals spent an average of 11 percent (206 seconds) of the time foraging from the boards. The boards were used by the animals with consistency; there was no indication that they lost interest in them over time.

Fekete et al. (2000) mounted a 15 x 41 cm large turf board inside, on a shelf of the cages of 10 pair-housed adult female squirrel monkeys and sprinkled a mixture of nuts, seeds and dried fruits onto the board on 11 consecutive days, right after the normal food was distributed. During the first 20 minutes, the animals spent on average 36 percent (7.3 minutes) of the time foraging from the board and ingesting the food they retrieved.

Chamove and Scott (2005) placed a 29 x 13 x 12.5 cm large *forage box* filled with a mixture of sawdust and food items into the cages of four female and four male individually housed adult marmosets, several hours before the daily standard food ration was distributed in open bowls. Over a 13-day test period, the monkeys spent 13 to 70 percent of the first hour searching for and retrieving food items from this box.

4.1.2.4. Access to the Vertical Dimension

The biologically inherent need of nonhuman primates to access the safe arboreal dimension can be met in the laboratory setting by installing resting surfaces in the animals' primary enclosures, preferentially at a height that allows the animals to retreat above eye level of humans (International Primatological Society, 1993; National Research Council, 1998; European Commission, 2002) who, after all, are natural predators for them (**Figure 41**).

Commercial built-in perches are often placed at such a low height, i.e., less than 30 cm (e.g., Bryant et al.,1988, Figure 1A; Watson, 2002, Figure 1; Reinhardt, 2003a, Figure 1; Allentown Caging Equipment, 2002), that they no longer serve the intended purpose of providing the occupant(s) access to the vertical dimension, but rather block part of the minimum floor area that is required by the animal(s) to turn around freely without touching the perch and the side walls of the cage (**Figure 42**).

In order to be useful, the resting surface (e.g., a perch or a platform) should reach from the back to the front of the cage so that an animal can:
1. freely move or sit under it (**Figure 43**),
2. retreat on it to the back of the cage during alarming situations,
3. sit on it in the front of the cage and maintain visual contact with other animals in the room (**Figures 41 & 43**).

Clarence et al. (2006) observed four pair-housed adult female rhesus macaques who lived in 280 cm high cages, each equipped with

Figure 41. A high perch offers caged primates a species-appropriate resting surface. Note that this male rhesus macaque shows no signs of distress while being approached by personnel; he seems to be free of anxiety or fear.

Figure 42. Commercial built-in perches are often placed much too low, thereby blocking part of the minimum floor area that would be required by an animal—here, two individually caged baboons—to turn around freely.

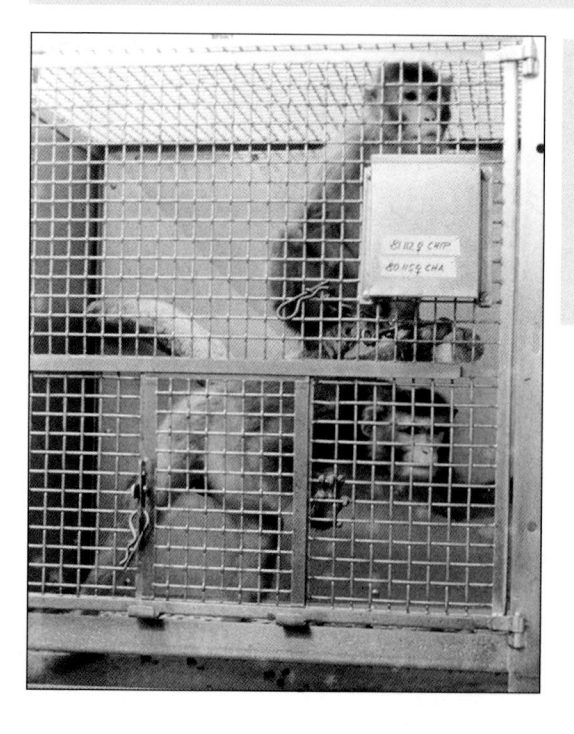

Figure 43. In the standard-size cage, the perch should be placed in such a way that an animal can freely turn around under it and sit on it at the front or at the back of the cage.

two same-sized platforms, one mounted 200 cm, the other 140 cm above the woodchip-covered ground. During 20 half-hour sessions, the animals spent on average:
- 70 percent of the time on the high platform, versus only
- 4 percent of the time on the lower platform or on the ground.

Reinhardt (1990a) tested 60 pair-housed rhesus macaques who had lived for 18 months in upper-row tier standard double-cage units each furnished with two perches. The perches consisted of gray, 10 cm-diameter polyvinyl chloride (PVC) pipes that were suspended diagonally with a slope of about 15 degrees. The lower end of a pipe was attached with a chain at the front of the cage, 175 cm off the ground, while the upper end rested at the junction of the back and side wall at a height of 185 cm (Reinhardt, 1989b; Reinhardt, 1990b). During one-hour observation sessions, the perches were used on average:

Figure 44. Animals caged in the bottom row live much closer to the ground and in a much darker environment than animals caged in the top row. A properly installed perch enables them to sit at least a little bit higher and at a shorter distance to the light source.

- 8 percent of the time by 42 adult animals (9 to 30 years old), and
- 18 percent of the time by 18 sub-adult animals (3.5 to 4 years old).

Access to an elevated surface seems to be particularly important in the traditional double-tier caging system for animals who are caged in the bottom rack. Living close to the "unsafe" ground in the shade of the upper row, these animals receive very little light (**Figure 44**). Access to the vertical dimension exposes them to more light and presumably enhances their feeling of security, as they can rest at a greater distance from the ground.

Woodbeck and Reinhardt (1991) compared perch use of 28 adult female rhesus macaques who lived since two years in double-cages located either in the upper row 140 cm above the ground (n=14) or in the lower row 30 cm above the ground (n=14; **Figure 44**). Each cage was furnished identically with two 10 cm-diameter PVC pipes.

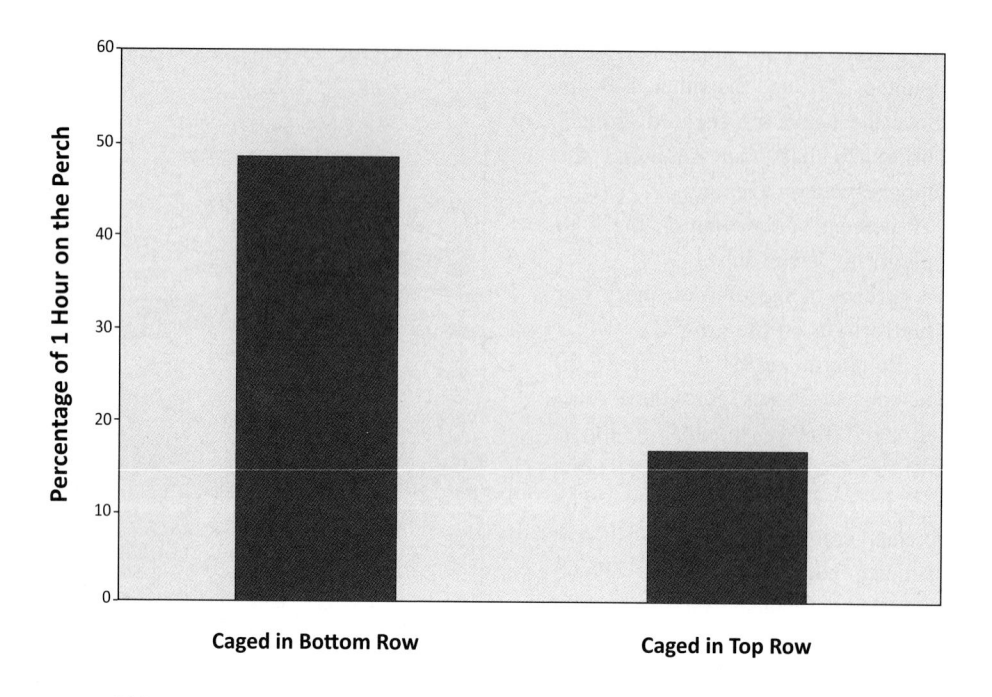

Figure 45. Mean time spent perching by 25 single-caged adult male rhesus macaques, caged either in the bottom row (n=11) or in the top row (n=14) of the cage rack (Reinhardt, 1989b).

During 30-minute test sessions, the monkeys sat on a perch on average:
- 7 percent of the time when they were living in the upper row, versus
- 32 percent of the time when they were living in the lower row; the difference was statistically significant. Animals in the bottom row probably have a greater need to sit above the floor of their cages because it makes them feel safer and exposes them to more light **(Figure 44).**

While perching, the animals were located:
- 74 percent of the time in the front of the cage,
- 26 percent of the time in the middle and back of the cage.

Reinhardt (1989) confirmed these findings in adult rhesus males who had lived for one year alone in upper-row (n=14) or lower-row cages (n=11), each furnished with a diagonally suspended 10 cm-diameter PVC pipe. During two one-hour observations, individuals caged in the bottom row sat on their perch for a significantly longer time than those caged in the top row **(Figure 45).**

While perching, the animals sat in front of the cage 95 percent of the time, and in the middle and back of the cage 5 percent of the time.

Bayne et al. (1992b) observed eight adult rhesus males during eight 30-minute

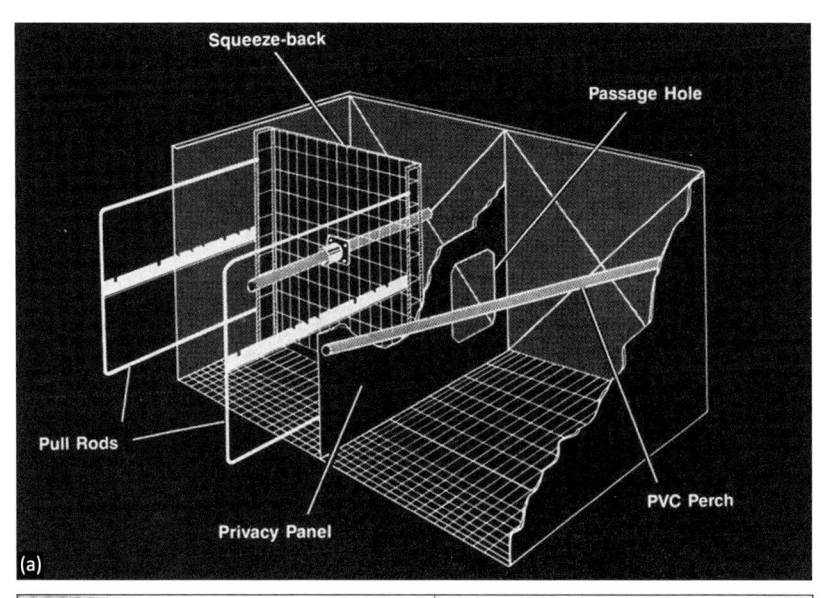

Figure 46a,b. This double-cage is equipped with two perches, one squeeze-back and a privacy panel. Note that the perch in the left half of the cage does not interfere with the operation of the squeeze-back.

sessions. The animals were kept individually [presumably on the upper row] in cages that were each furnished with three enrichment devices and one galvanized steel perch of unspecified diameter, which was *placed approximately 20 cm off the floor of the cage, parallel to the side wall*. The males sat on their perches on average 17 percent of the time.

When rhesus macaques are given the choice of sitting in one half of a double-cage on a perch made of PVC or wood, they show no significant preference for either material (Reinhardt, 1990c).

Elevated resting surfaces can readily be installed in standard cages. Schmidt et al. (1989a) and Reinhardt and Pape (1991) developed two different designs for cages with squeeze-back walls (**Figure 41**). In both instances, the perch (a) runs parallel to the sides of the cage, allowing an animal to sit in the back or in the front of the cage, and (b) allows the squeeze-back mechanism to slide freely from the back to the front of the cage. The diameter of the perch is predetermined by the bar spacing or the wire mesh size of the squeeze-back in the design by Schmidt et al. (1989a)—typically about 2 cm—but not in the design by Reinhardt and Pape (1991)—typically about 10 cm (**Figure 46a,b**). Kenney et al. (2006) developed a perch that automatically folds flat against the side wall of the cage and can be pulled down by the animal(s) to a horizontal position, providing a ledge on which to sit or stand. The squeeze-back has to be adjusted so that it can be moved over the folded perch.

While caged macaques make use of and benefit from fixed elevated resting surfaces such as perches and platforms, they show little interest in swings (Dexter and Bayne, 1994). The spatial constraint of the standard cage does not allow for true swinging. When they have the choice, adult rhesus will clearly prefer sitting on a PVC pipe that is mounted onto the back and front walls rather than suspended from the ceiling of the cage (Kopecky and Reinhardt, 1991). It is probably more comfortable for a monkey to rest on a stable rather than unstable raised structure.

4.1.2.5. Environmental Enrichment

Environmental enrichment temporarily enhances well-being if it provides opportunities for the expression of behaviors that have survival value, such as foraging and retreating to the vertical dimension. There is very little evidence that environmental enrichment also helps the confined subject to cope with permanent confinement distress as reflected in serious behavioral pathologies.

It has been claimed repeatedly that self-biting and hair-pulling can be controlled to some extent with environmental enrichment (Bryant et al., 1988; Gilbert and Wrenshall, 1989; Erwin, 1991; Watson, 1992; Watson et al., 1993; Niemeyer et al., 1996; Tustin et al., 1996; Storey et al., 2000; Marshall et al., 2002; Turner and Grantham, 2002; Tully, 2003; Honess et al., 2005), but there is only one report to support this claim with scientific data. Smith et al. (2004) describe the case of an adolescent female chimpanzee who engaged in hair-pulling to the point of creating open lesions. The animal was offered large quantities of shredded paper to add opportunities for non-self-directed activities. Systematic behavioral data were collected for a 10-day period prior to the provision of enrichment, and for a three-month period during which the animal had uninterrupted access to paper. Hair-pulling decreased already on the first day when the animal received shredded paper and it continued to decrease with prolonged exposure. The chimpanzee used the paper in different ways; one of them resembled *leaf-pile pulling*, a behavior pattern reported in wild chimpanzees (Nishida and Wallauer, 2003).

4.2. Separation from the Companion

Separation from and loss of a companion is a major stressor for human primates (Biondi and Picardi, 1996; Hamiel et al., 1999; Shear and Shair, 2005); there is good reason to believe that the same holds true for nonhuman primates, who, like humans, develop strong, long-lasting bonds with each other (Chance, 1956; Chance, 1961; Chance and Jolly, 1970; Chance, 1975; de Waal and Luttrell, 1986; Fruth and Hohmann, 1998; Casanova and Garcia, 1996; Hemelrijk et al., 1999; Stopka et al., 2001; Silk, 2003; Fujisawa et al., 2004; Hermano-Silva and Lee, 2004; Smuts, 2004; Bonnie and de Waal, 2006; Duffy, 2006; Kapsalis and Johnson, 2006; Silk et al., 2006; Nakamichi and Yamada, 2007; Shibata and Ford, 2007; Watts, 2007).

4.2.1. Signs of Distress and Impaired Well-Being

Being forcefully separated from the companion is an intrinsic stressor that is reflected in behavioral, vocal, endocrinological and cardiovascular stress responses (Rasmussen, 1985; Hennessy, 1997; Smith and French, 1997; Watson et al., 1998; Gerber et al., 2002; McMillan et al., 2004), and subjects can be so traumatized that they react by injuriously biting themselves (Maple et al., 1973; Anonymous, 2004).

4.2.2. Alternatives to Partner Separation

There are three situations in which pair-housed animals are typically separated because it is believed—but not proven—that the presence of another conspecific would jeopardize an animal's safety and interfere with data collection and research protocol.

4.2.2.1. Post Operative Recovery

Murray et al. (2002) challenged conventional wisdom and allowed 15 pair-housed female long-tailed macaques to return to their companions on the same day of vascular access port surgery once they had fully recovered from anesthesia. Change in hierarchy status, self-traumatic events, weight loss or diarrhea did not occur in any of these animals, and the incision sites healed without complication. The animals ate and drank normally and readily accepted post-operative oral medication.

Baumans et al. (2007) cite a report on a long-tailed macaque colony in which 95 percent of the animals are pair-housed:

> *The animals are subjected to a lot of orthopedic procedures. There have never been problems with the re-pairing of the animals after surgery. We partition the pair's cage with a transparent panel, which we remove after the treated companion has fully recovered from anesthetic effects (usually 24 hours). It has never happened that animals who had no surgery showed any negative behavioral reactions toward their temporarily probably weaker cage mates.*
>
> *In a small study we compared post-op recovery of the animals when:*
>
> *a) only one partner had surgery resulting in a full length cast on one of the legs,*
>
> *b) both companions had the surgery, and*
>
> *c) the animal, who had surgery, was kept alone for a few days.*
>
> *We found that there was:*
>
> *• less cast picking,*
>
> *• faster recovery, and*
>
> *• quicker return to full range of motion after the cast had come off, when the animals were re-paired with their partners, than when they were kept alone after surgery.*

4.2.2.2. Food Intake and Metabolic Studies

Reinhardt and Reinhardt (2001) install wire mesh partitions prior to food distribution. In this way, paired partners are separated in their familiar homecages, but maintain visual, olfactory and auditory contact while one or both of them are being tested (**Figure 47a, b**). After food intake for the day has been recorded, the dividing panel is pulled so that the two animals have full contact with each other during the night until new food is distributed the next morning.

A wire mesh divider is also an option for studies requiring the collection of urine and feces. It allows cage companions to keep uninterrupted contact with each other without interfering with the collection of individual-specific urine and feces samples.

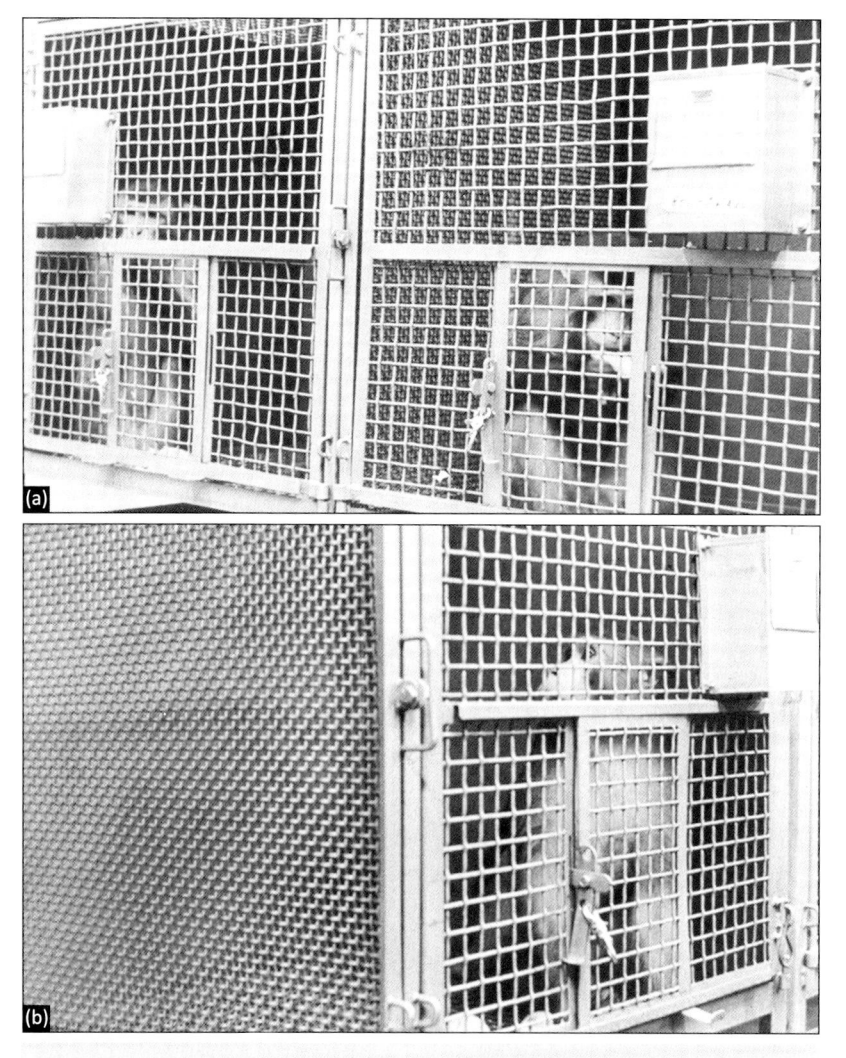

Figure 47a,b. For food-intake studies, paired rhesus macaques *Klaus* and *Mark* are separated in their home cage (a) with a grated cage divider (b) that is removed during the night when food intake is not assessed.

4.2.2.3. Neurophysiological Studies

It has been repeatedly documented that keeping compatible pairs of rhesus macaques together, after one or both partners have been instrumented with cranial implants, does not jeopardize the safety of the animals and the safety of the implants, and also does not interfere with physiological testing (**Figure 48a,b**; Reinhardt, et al., 1989; Reinhardt and Dodsworth,1989; Reinhardt and Reinhardt, 2002). Roberts

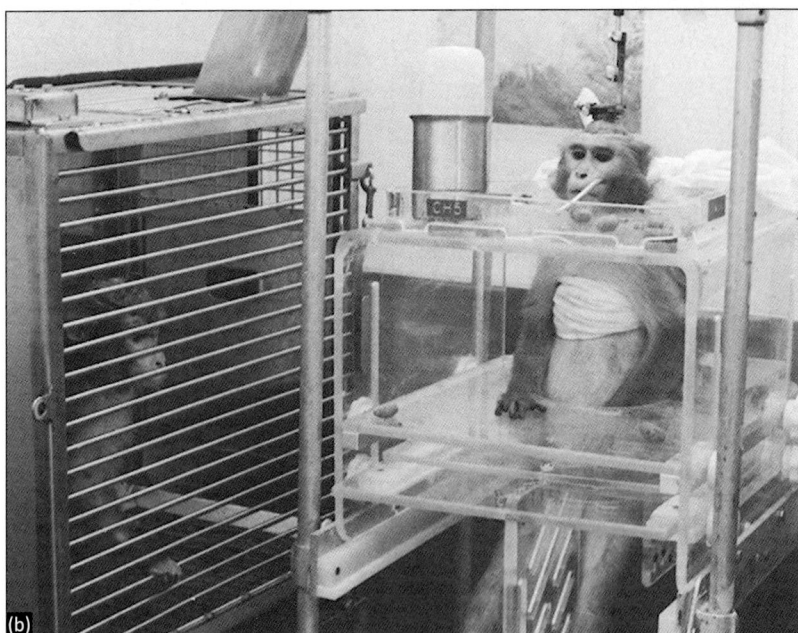

Figure 48a,b. Pair-housed rhesus macaques *Gina* and *Sylvia* with cranial implants in their home cage (a) and during experimentation, when one partner is chair-restrained, while the other partner provides psychological support in a mobile cage (b).

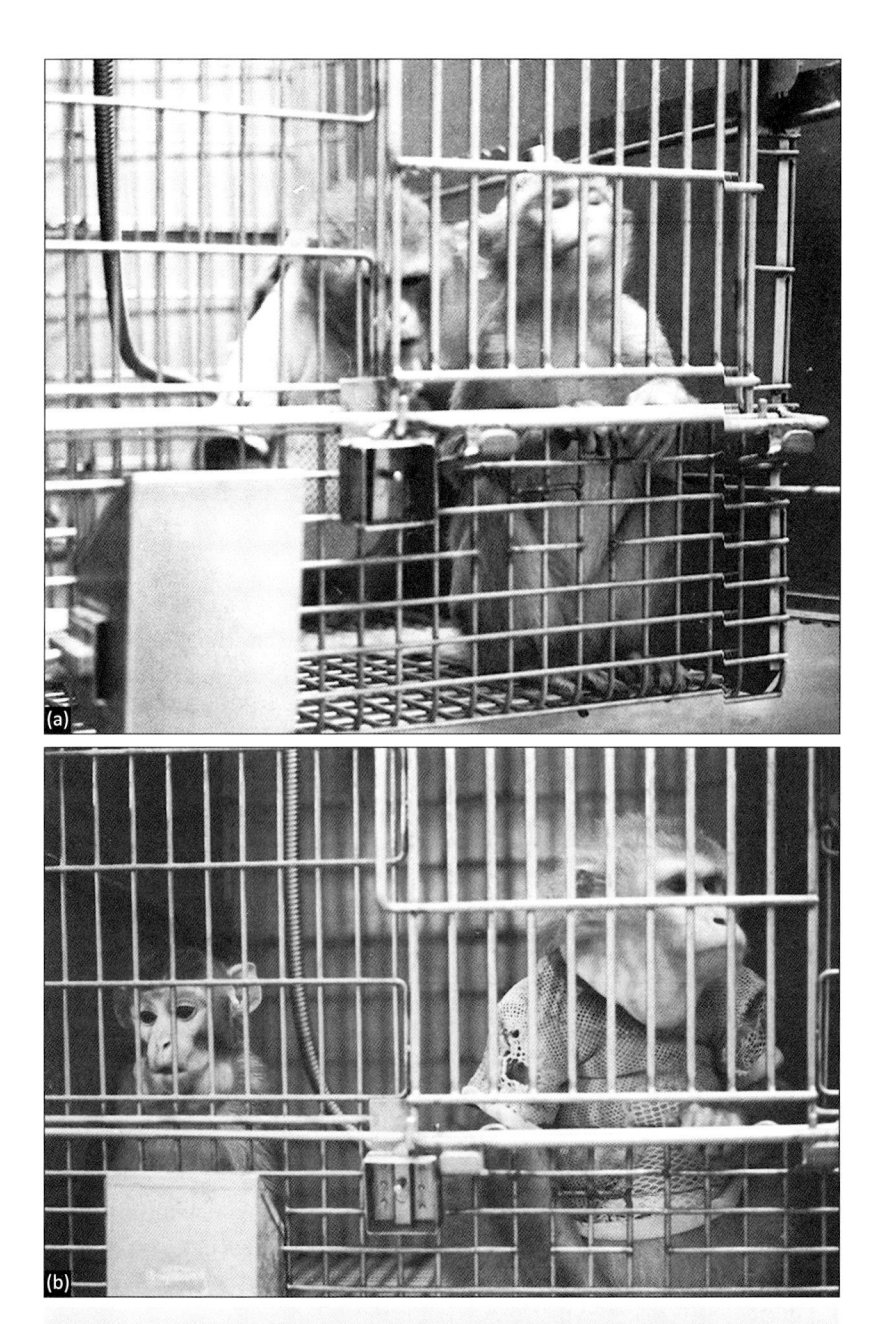

Figure 49a,b. Tethered rhesus macaques *Betty* (a) and *Chewy* (b) with their juvenile companions *Lissy* and *Cute* during an experiment requiring remote sample collection. Note that *Betty* grooms *Lissy,* who has a cranial implant (a).

and Platt (2005) confirmed these clinical observations in six cranial-implanted adult male rhesus macaques who lived for several years in compatible pair-housing arrangements without adverse effects on their clinical health and without adverse effects on the implants.

Paired animals are also regularly separated when one or both of them are assigned to physiological studies requiring remote sample collection via a tether system.

Coelho and Carey (1990) designed a *social-tether cage system* for baboons that gives tethered cage neighbors tactile contact with each other through grated dividing panels. This system provides an advantage in that:

> *Socially housed baboons interact with compatible cage neighbors, while individually housed baboons attempt to shake and dismantle their cages. During the four years that the social-tether cage system was used with several hundred baboons, it never happened that neighboring baboons bit the hand or fingers of each other and they never pulled the catheter or attempted to remove or dismantle the jacket of another animal.*

In some cases, there may actually be no need to separate partners with a wire mesh panel when one of them is tethered: Reinhardt (1991c) and Reinhardt (1997) documented two cases of adult-infant rhesus macaque pairs in which the presence of the young companion did not interfere with the tethering of the adult companion for remote sample collection (**Figure 49a,b**).

4.3. Social Conflicts

Conflicts among otherwise compatible social partners are unavoidable. In the wild, they are relatively rare and subtle because the animals have the necessary space to get away from each other as dictated by dominance-subordinance relationships (Hall and De Vore, 1965; Southwick et al., 1965; Kummer, 1968; Van Lawick-Goodall, 1968; Chance and Jolly, 1970; Wheatley, 1999).

4.3.1. Signs of Distress

The unnatural spatial restrictions in the research lab setting does not allow nonhuman primates to maintain inter-individual social distances as needed. Overt aggressive conflicts can, therefore, be quite common. Individuals may become the target of repeated overt aggression from their cage companions. This will make them extremely anxious, intimidated and depressed, a situation that finally necessitates the separation of the two animals (**Figure 50**).

Social distress is also often caused when an animal is transferred to a new housing area in which the residents constantly intimidate the newcomer (**Figure 51a,b**).

DISTRESSING CONDITIONS

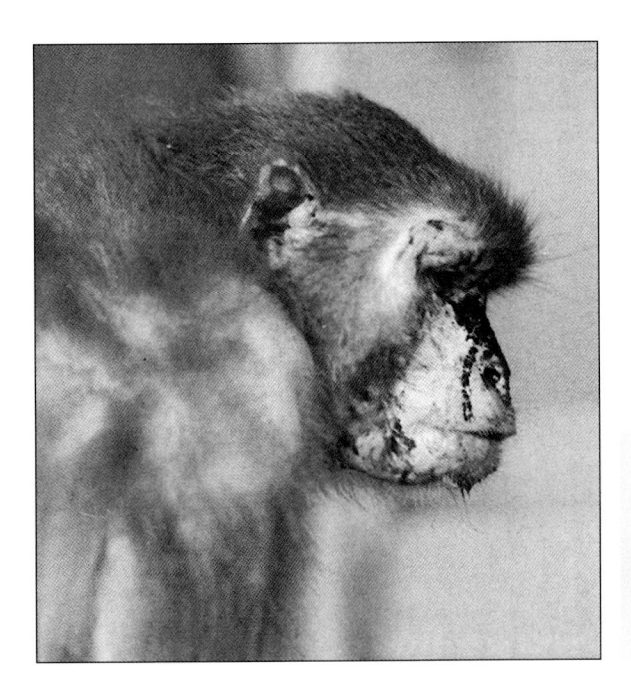

Figure 50. Adult rhesus macaque *Eve* is depressed because she has been repeatedly harassed by her cage mate.

Figure 51a,b. Rhesus macaque *Kim* has been moved to a new room (a) where she is constantly threatened by animals from across the aisle (b).

Figure 52a,b. With a privacy panel, paired rhesus males *Moon* and *Grey* spend most of the time in the same half of the cage (a) (Reinhardt and Reinhardt, 1991; Basile et al., 2007), but they can break visual contact, especially when they collect biscuits from the food boxes (b).

4.3.2. Refinement

4.3.2.1. Breaking Visual Contact

Reinhardt and Reinhardt (1991) designed *privacy panels* for 30 adult pair-housed female rhesus macaques: A sheet of stainless steel with a passage hole divides the double-cage in such a way that the two partners have the option of accessing one of the two food boxes in a different half of the cage without being seen by each other (**Figure 52a,b**). With the privacy panels in place:
1. Dominant partners no longer tried to prevent their subordinate cage mates from getting food.
2. Companions spent more time grooming and hugging each other.
3. The incidence of conflicts—expressed in fear-grinning, threatening, pushing and slapping—decreased.

As a consequence of these results, privacy panels were installed throughout the colony of more than 600 pair-housed rhesus macaques. Basile et al. (2007) concluded from similar findings that *a privacy divider may provide a safe haven and give monkeys the ability to diffuse hostile situations before they escalate.*

Ratajeski and McDonald (2005) mention a case study in which a sub-adult female long-tailed macaque pulled large amounts of hair from her caudal area and posterior thigh following relocation to a new housing room. The animal was obviously very intimidated by her new neighbors and spent much of the time clinging to the upper back wall of her cage (**Figure 51a**). To alleviate the distress, a blind was installed so that the newcomer could choose to avoid visual contact with other animals in the room. This had the effect that *the female's hair-pulling and clinging behavior **ceased*** [emphasis added].

4.3.2.2. Access to the Vertical Dimension of the Enclosure

Kitchen and Martin (1996) observed five adult female-male pairs of marmosets for 20 hours distributed over 12 days (a) in their standard home cage without furniture, and (b) in their home cage equipped with three perches. Access to the perches resulted in a significant decrease in aggression (**Figure 53**).

Access to elevated structures is likely to moderate aggression also in other primate species as it has been shown that the provision of high perches significantly decreases aggression among group-housed mangabeys (Neveu and Deputte, 1996) and Japanese macaques (Nakamichi and Asanuma, 1998).

Figure 53. Access to elevated structures helps marmosets diffuse social tensions by allowing cage mates to quickly increase social distance as needed.

4.3.2.3. Careful Re-Introduction after Separation

Overt aggression among compatible cage mates is often unintentionally provoked when they are reunited after one of them has been separated for research-related reasons; the two animals don't recognize each other instantaneously and, therefore, treat each other as strangers and start fighting.

Empirical evidence indicates that this risk can be avoided by giving temporarily separated partners the chance to recognize each other first and then reunite them. This can be accomplished by partitioning the pair's home cage with a transparent panel, and then introduce the partner who had been away into the empty section of the cage. The two companions will quickly recognize each other and treat each other accordingly when the dividing panel is removed (Reinhardt, 1992a; Jackson, 2001).

4.4. Enforced Restraint

Restraint during clinical procedures and sample collection is a distressing experience not only for human primates (**Figure 54a**; Selekman and Snyder, 1996; Tomlinson, 2004; Folkes, 2005; Melhuish and Payne, 2006; Bland et al., 2007; Brenner, 2007) but also for nonhuman primates, who unlike humans are usually restrained with force without their consent (**Figure 54b**).

Published information provides scientific evidence that traditional, involuntary restraint techniques of research non-human primates are intrinsically a source of distress resulting from fear (Reinhardt et al., 1995, p 221). Research data collected from a distressed monkey are "distressed" and hence of little scientific value (Reinhardt, 1998, p 18). There is no scientific evidence that the animals adequately habituate to involuntary restraint (Reinhardt et al., 1995, p 221). Physical restraint procedures should be used on awake animals only after alternative procedures have been considered and found to be inadequate. If a restraint will be utilized the animal should be trained or conditioned to the restraining device, using positive reinforcement, prior to the beginning of the experiment (Prentice et al., 1986).

4.4.1. Signs of Distress

Handling practices of primates traditionally bear two serious stressors for the individual subject:

1. *Being forcefully caught and removed from the home cage* triggers behavioral distress responses and significant endocrinological and cardiovascular stress reactions (Mitchell and Gomber, 1976; Phoenix and Chambers, 1984; Herndon et al., 1984; Line et al., 1987; Reinhardt et al. 1990b; Line et al., 1991; Crockett et al., 1995; Jorgensen et al., 1998; Gerber et al., 2002; Davenport et al., 2007).
2. *Being forcefully restrained* results in behavioral and emotional distress responses and significant hematological, endocrinological and cardiovascular stress reactions (Ives and Dack, 1956; Ackerley and Stones, 1969; Manning et al., 1969; Berendt and Williams, 1971; Quadri et al., 1978; Goosen et al., 1984; Golub and Anderson, 1986; Wheeler et al. 1990; Line et al., 1991; Brockway et al., 1993; Schnell and Wood, 1993; Fowler, 1995; Klein and Murray, 1995; Reinhardt and Reinhardt, 2001).

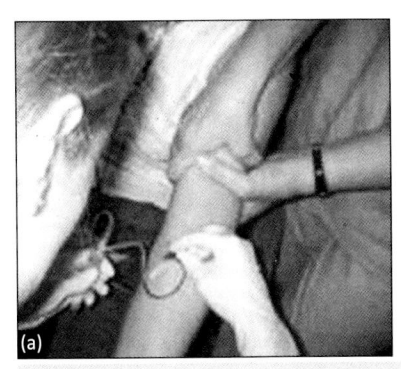

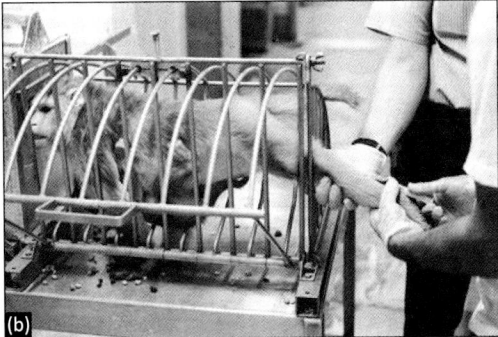

Figure 54a,b. Restraint during unpleasant procedures can be a distressing experience for human primates (a) and nonhuman primates alike (b).

There are numerous scientific articles mentioning that caged primates can be trained or were trained to cooperate during common procedures in order to reduce or eliminate data-biasing stress reactions (Michael et al., 1974; Elvidge et al., 1976; Byrd, 1977; Rosenblum and Coulston, 1981; Herndon et al., 1984; Wall et al., 1985; Whitney and Wickings, 1987; Jaeckel, 1988; Suleman et al., 1988; Hein et al., 1989; Scallet et al., 1989; Chambers et al., 1992; Reichard and Shellaberger, 1992; Eaton et al., 1994; Hernándes-López et al., 1998; Hrapkiewicz et al., 1998; Nelms et al., 2001; Bentson et al., 2003; Grant and Doudet, 2003; Iliff et al., 2004; Koban et al., 2005). There are only a few reports describing and evaluating the techniques used to achieve the goal of such training.

4.4.2. Refinement

4.4.2.1. Training to Cooperate during Injection and Venipuncture

Levison et al. (1964) developed a *technique by which a large, aggressive male* **baboon** *was trained to offer his arm to receive an injection,* rather than being forcefully chair-restrained during this routine procedure. The front wall of the baboon's cage contained a 9 cm-diameter porthole.

> *The training procedure was begun by holding a slice of fruit in front of the hole and giving it to the male when he extended his arm through the opening. Then, the fruit was given only when the arm was fully extended, and later, held quietly for a number of seconds.*
>
> *On the trials that followed, the baboon was required to maintain this behavior while the experimenter touched his arm in a progression of closer approximation to drug injection. The baboon was given fruit after each satisfactory extension. Reinforcement was withheld if the wrong arm was extended, or if the arm was bent or withdrawn in any degree in response to tactile stimulation. The trainer:*
> 1. *touched, and later held the baboon's wrist with his left hand;*
> 2. *touched the biceps with his right hand, and then with the syringe, while holding the animal's wrist firmly;*
> 3. *made injection contacts in which the syringe and needle were placed against the arm and finally inserted into the muscle.*
>
> *Only two training sessions were required before the needle could be held against the animal's biceps. Emotional displays and withdrawal of the arm occurred more frequently after the first penetration of the needle; however, the behavior was brought well under control when a special procedure for inserting the needle was begun.*

Figure 55. This originally aggressive male baboon has been successfully trained to voluntarily present his arm for test drug injection in his home cage. Note that the male is not forced with a squeeze-back to tolerate the procedure.

The experimenter would press down progressively harder on the biceps muscle with the side of the needle, then slowly slide the point forward into the muscle while maintaining the lateral pressure. The point of the needle was not in contact with the skin until the forward move to insert it was made. After insertion, the needle was held in the muscle for successively longer periods; then, an actual injection was performed.

Satisfactory injection was reliably obtained after approximately three weeks of one hour-training sessions on alternate days. The baboon continued to cooperate when both active and control compounds were injected by two different researchers (**Figure 55**).

Priest (1990, 1991a) provides a detailed description and video-document about how he trained an adult single-caged diabetic **drill** (*Mandrillus* sp.) to cooperate during insulin injection and blood collection in the subject's home cage:

Because of Loon's medical condition, our first training priority was to condition him to accept his insulin injections voluntarily. This was begun in July, 1989, at the Zoo's veterinary hospital by hospital technicians. Necessary daily injections were being administered using a squeeze cage. By simply pairing a food reward with his daily injection, we began to establish the medical procedure as a positive event. In the early stages of conditioning, it was necessary to continue to use the squeeze cage to immobilize him. However, Loon quickly learned to recognize the injection as a precursor to food. By pairing his afternoon meal with an injection, while at the same time fading the use of the squeeze cage, the need for immobilization quickly became unnecessary. Within a few days, Loon learned to offer his back for

---

[1]Retouched by Annie Reinhardt; reproduced with permission from Levison PK, Fester CB, Nieman WH and Findley JD 1964 A method for training unrestrained primates to receive drug injection. *Journal of the Experimental Analysis of Behavior* 7: 253-254; Copyright 1964 by the Society for the Experimental Analysis of Behavior, Inc.

the injection in anticipation of the reward. In addition to the food reward, Loon was being positively reinforced by the physical freedom made possible by his compliance.

Our [next] priority was to train him to allow venipuncture for blood sampling. Loon was trained to reach into a stainless steel tube, cut to the exact length of his arm, and to grasp a steel rod positioned crosswise at the end of the tube. As long as the drill was grasping the rod, he could not easily grab the trainer. Within three days of his exposure to a formal program of operant conditioning, Loon was grasping the rod and holding this position until a bridging stimulus (a clicker) was sounded, signaling termination of the behavior and presentation of a food reward.

Through an ellipse cut in the tube, I began to desensitize the drill to touch on his shaved forearm while he grasped the rod at the end of the tube. I began by reinforcing his allowing me to groom his arm and, on a separate command, his back. In addition to the social rewards baboons attach to grooming, Loon was also being rewarded with food items.

As training progressed I would occasionally drag different items over the bare skin of his forearm. This procedure desensitized him to a variety of stimuli, and simultaneously provided an occasion to reward him for grasping the rod.

During the first several weeks of training, Loon was very aggressive. He would snatch the food reward and, if I were not quick enough in removing my hand, take a swipe at me. On several occasions he succeeded in tearing the surgical glove off my hand. To reduce his aggression, we rewarded him with additional treats when he took the reward gently.

About six weeks into his training, Loon's medical condition required a blood sample. He was given the command to place his arm in the tube and grasp the rod. Within moments, a veterinarian had withdrawn the blood sample. Loon continued to wait patiently for the bridging stimulus to terminate rod-holding. The blood withdrawal had apparently been of no concern to him as he focused on holding the rod.

As a result of the need to test Loon's blood frequently, the veins in both of Loon's forearms have become heavily scarred. Loon has tolerated up to six failed attempts to draw blood from these battered vessels, without ever once pulling his arm away from the tube and rod. We responded to this new problem by training Loon to offer the vessels on the ventral side of both of his legs for venipuncture. Now venipuncture sites are rotated to help reduce damage to any single vessel site.

In nearly one year of training, Loon has never failed to voluntarily accept his insulin injection or to allow the veterinarians access to blood vessels in exchange for a good back scratch and a food reward (Priest, 1991b).

Laule and Whittaker (2001), Schapiro (2005) and Pranger et al. (2006) applied the

same venipuncture training technique successfully with adult **chimpanzees** (**Figure 56**) living in pairs and small groups, and adult individually housed rhesus macaques of unspecified gender. It took an average of:
- 219 minutes in 31 sessions to successfully train four chimpanzees, and
- 156 minutes in 32 sessions to successfully train two macaques.

McGinnis and Kraemer (1979) and Laule et al. (1996) used a less protective positive reinforcement training technique to obtain cooperation of adolescent female chimpanzees. While McGinnis and Kraemer (1979) document their success with a photo (**Figure 57**), Laule et (1996) describe their training technique:

Figure 56. The "protected" blood collection training technique, originally developed by Priest (1990), is here successfully applied with a chimpanzee who is rewarded with fruit juice for cooperating during blood collection.

Allie was nursery-raised and, hence, extremely tractable prior to the onset of the formal training, which initially implied that she had to sit upright and allow her arm to be manipulated and held by the trainer.

Next, she was desensitized to having her arm touched by, first, the trainer's finger, then a cotton swab, and then a syringe without a needle, with a blunt needle, and finally with a sharp needle. Throughout the process Allie was rewarded for being calm and for tolerating each stimulus for increasingly longer periods of time.

The first attempt to actually draw blood occurred during the 18th training session, with a total of 275 minutes of training time invested prior to that. The attempt was successful; Allie showed no visible signs of stress or discomfort, sat quietly, watched the entire procedure, and eagerly accepted rewards. During subsequent blood draws, she has never refused or disrupted the procedure.

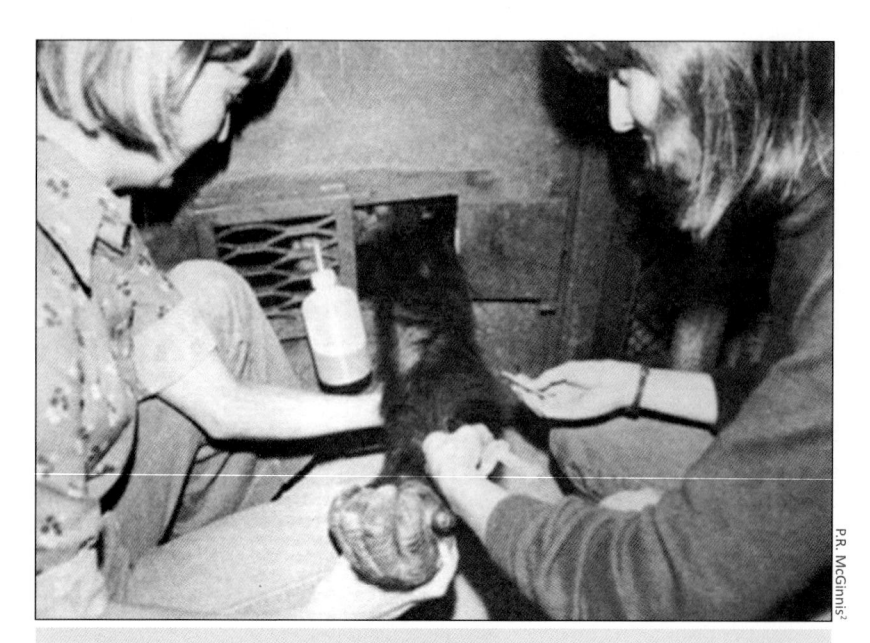

Figure 57. Adolescent chimpanzee Joe is rewarded with apple juice for his cooperation during blood collection.

Reinhardt and Cowley (1992) worked with six adult female **stump-tailed macaques** who were pair-housed for more than one year in double-cages, each provided with a privacy panel, two perches, and one restraint mechanism. The animals were accustomed to being restrained with the squeeze-back for husbandry-related procedures. The door of the restraint compartment was equipped with a sliding transparent Plexiglas panel. Its opening allowed an animal to comfortably extend a leg out, yet was small enough to prevent the animal from protruding the head out of the cage (**Figure 58a-d**). The panel was also used as a safeguard for the person performing the venipuncture. An animal could be trapped by pulling the squeeze-back past the passage hole of the privacy panel (**Figure 46a**). The companion had free access to the rear portion of the squeeze-back, allowing visual contact.

The animals were used to having blood collected in a restraint apparatus away from their home cages. They were familiar with the authors who trained four and two of them, respectively. The training protocol comprised the following steps:
1. The subject is enticed with favored food to enter the restraint compartment of the double-cage.
2. By pulling the rods of the squeeze-back, the subject is restricted to the front quarter

[2] Reproduced with permission from *Comfortable Quarters for Laboratory Animals* (Seventh Edition), 20-27, Animal Welfare Institute, Washington, DC, 1979.

Figure 58a-d. Adult stump-tailed macaque *Zora* has been trained to voluntarily present a leg for blood collection in her familiar home cage (a-c). She is praised with "Good Girl!" and rewarded with raisins, her favored treats (d).

of the restraint compartment. This restricts her freedom of movement but still allows her to turn round and climb up the mesh walls of the cage. The animal is gently scratched through the mesh and food-rewarded.
3. The spatially restricted subject is enticed with food to face the left or right side of the cage. Her back is gently scratched through the opening of the Plexiglas panel. This again is followed by a food reward.
4. The subject's back and thighs are scratched. One of her legs is gently lifted and firmly pulled toward the opening of the panel. A food reward follows.
5. The subject's leg is pulled through the opening of the panel and a blood sample taken by means of saphenous venipuncture. The procedure is again concluded with a food reward.
6. Once the subject passively tolerates the above procedure with no signs of resistance, she is restrained in one third of the compartment, rather than one quarter, thus allowing free movement. Venipuncture is carried out and the animal rewarded. This exercise is repeated on different occasions until the animal spontaneously cooperates (**Figure 58a,b**).
7. Restrained in one third of the cage, the subject actively cooperates, i.e., voluntarily presents a leg behind, or through, the opening of the Plexiglas panel and accepts venipuncture (**Figure 58c**); this is followed by a food reward and praise (**Figure 58d**).

All six stump-tailed macaques were successfully trained within a two-week period to actively cooperate during blood collection in their home cages. Nine to 23 training

Animal	TRAINING TIME TO ENSURE:			SERUM CORTISOL AT:	
	passive tolerance (min)	active cooperation (min)	total training (min)	13:00 (µg/dl)	13:15 (µg/dl)
Jean	8	7	15	28.0	28.6
Einstein	17	12	29	28.4	28.2
Browny	22	10	32	24.3	21.4
Agy	17	20	37	31.0	31.1
Lucy	14	29	43	31.9	36.6
Goldy	20	25	45	26.7	24.9
Mean	**16.3**	**17.2**	**33.5**	**28.4**	**28.5**

Table 1. Cortisol response to volunatry blood collection and time investment to train pair-housed adult female stump-tailed macaques to cooperate during blood collection in their home cages.

sessions per monkey were necessary to achieve this. Sessions were scheduled according to a subject's progress, although individuals were trained on no more than three occasions per day. The monkeys were trained with firm gentleness, but no sessions were terminated before the goal of the training step was achieved. The duration of individual sessions was therefore not constant but varied between 49 and 351 seconds. Animals who resisted the conditioning process (e.g., were unwilling to turn to one side, climbed up the cage wall to avoid having the leg grasped, struggled while having the leg pulled out of the cage) were never punished but treated with special patience.

On average, 16 minutes of training time was invested until the monkeys *passively* tolerated in-homecage venipuncture (steps 1-5). An additional 18 minutes were then required to ensure *active* cooperation during the procedure (steps 6 and 7). Total average training time was thus 34 minutes, ranging from 15 to 45 minutes (**Table 1**). It is sometimes argued that the training of nonhuman primates to cooperate during procedures has the *disadvantage* of requiring *considerable* time to be executed successfully (Klein and Murray, 1995; Hrapkiewicz et al., 1998). The investment of less than one hour per animal suggests that this does not hold true in all cases and therefore should not discourage qualified animal care personnel to train primates in their charge.

Once trained, the six stump-tailed macaques no longer displayed behavioral signs of distress during blood collection: They did not resist and struggle and they did not try to scratch or bite the handler in self-defense. In order to evaluate possible physiological stress reactions, serum cortisol concentrations were measured. For this purpose, two 0.5 ml blood samples were collected from each animal one week after the last training session, at 13:00 and at 13:15. The subjects were undisturbed by human activity for 90 minutes prior to the first venipuncture at 13:00. The first sample was used to assess basal cortisol concentrations and the second to assess the magnitude of cortisol response 15 minutes after venipuncture.

Basal serum cortisol concentrations from the samples taken at 13:00 were not significantly different from those taken 15 minutes later, indicating that the animals experienced no stress while cooperating during blood collection (**Table 1**).

Reinhardt (1991d) and Reinhardt (2003b) applied this training technique with ten pair-housed adult male and 12 pair-housed adult female **rhesus macaques.** These animals were also used to being restrained with squeeze-backs for husbandry-related procedures and having blood samples taken under enforced manual or mechanical immobilization in a designated treatment area. On average:
- 20 minutes of training time was invested until the male subjects *passively* tolerated in-homecage venipuncture (steps 1 - 5); an additional
- 19 minutes were then required until the subjects *actively* cooperated during the procedure (steps 6 and 7; **Figure 59a-d**). Total training time ranged from
- 16 to 63 minutes, with a mean of 39 minutes.

Figure 59a-d. Rhesus macaque *Max* voluntarily presents a leg for blood collection while his companion *Ray* attentively watches (a,b). Cooperation is always reinforced with "Good Boy!" and a food reward (c). *Ray* is also rewarded because he has not disrupted the handling procedure that took place with *Max* (d).

Figure 60a-c. Rhesus macaque *Rocky* presents a leg and holds still during blood collection (a). She required little extra formal training to cooperate during injection (b). *Rocky* and also her cage mate *Tora*, who did not make any fuss during the procedure, are rewarded with grapes (b).

Average total training time for the females was the same as for the males, i.e., 39 minutes. This time investment does not seem unreasonably high when considering the long-term benefits of working *with* cooperative animals rather than *against* resisting animals. There were no behavioral indications that the trained animals experienced apprehension or fear during the blood collection procedure; all males cooperated not only with the trainer, but also with the attending care personnel, as well as with experienced personnel from other facilities.

Empirical experience has shown that animals who have been successfully trained to cooperate during venipuncture require hardly any extra formal training to obtain their cooperation also during intra muscular injection (**Figure 60a-c**).

Reinhardt (1992b) applied this training technique to six pair-housed juvenile (13 to 18 months old) female rhesus macaques. The training was successful in only one pair; the two juveniles required 46 and 47 minutes of training distributed over 38 and 37 sessions until they extended their legs through the cage opening for venipuncture. The training of the other four animals was discontinued after more than 40 sessions when it became clear that they were unduly distressed by being at such close quarters with a human "predator."

Stringfield and McNary (1998) successfully trained a *high-strung, suspicious, cautious* **red-tailed moustached guenon** (*Cercopithecus cephus cephus*) to accept daily insulin injection. *David* lived with two other monkeys. He was moved to a large squeeze-back cage to undergo training during two daily sessions. A clicker and a colored target were used, with food rewards being given for proper behavioral responses.

Within two months, *David* was expert at stationing and putting his arm through the bars to touch the target. He never became comfortable having his arm held or manipulated, and would retreat when his arm was handled. However, when he would approach in a less formal manner, it became apparent that he liked to present by lying down with his back facing the trainer. He would then allow his back and other parts of the body to be scratched. Training was adjusted accordingly and rapidly progressed within another two months from scratching his back, to pinching his skin, to poking with a needle, to injecting a small volume of saline, and finally to injecting insulin.

Bayrakci (2003) developed a technique to achieve active cooperation during injection from three individually housed, adult male **lion-tailed macaques** (*Macaca silenus*).

1. The first step in the training process was to help the monkeys recognize the clicker as an indicator of a correct response and an upcoming food reward. This was accomplished by calling the subject, "come here!" and then click while saying "good," and finally offering a food reward. While the animal was sitting attentively in front of the trainer, the trainer continued to click and food-reward. It took only a few sessions for the macaques to expect a reward after hearing the click, and a few more sessions to adjust to this relationship with the trainer.

2. Before starting injection training, a 5 cm-diameter hole was cut in the mesh wall 30 cm above the floor. Training sessions were conducted in front of this opening, so the macaques were comfortable sitting in front of it. The subject is shown a treat and rewarded for extending his arm outside the cage. The treat is then given through the mesh, not in front of the hole. The trainer quickly learned that if the macaque was rewarded through the hole, the arm extensions are too brief. When the subject reaches for the treat with one arm at a good distance away from the hole, it is easy to gradually increase the duration of the other arm's extension through the hole to allow enough time for an injection.

   The macaques were willing to extend their arms through the hole on command "Touch!" right from the beginning, so training sessions focused on increasing the length of arm outside the caging and the duration of that extension.
3. Once arm extension was established, the trainer added a bamboo stick poised above and to one side of the hole, and began to gently press on the arm when fully extended. The macaques rapidly got used to the stick and the trainer began to press harder. The stick was then replaced with an empty syringe without a needle, then with the plastic needle tip, then with a long blunted needle. The clicker and "good" followed by a food reward was used to reinforce full arm extension beyond the moment when the syringe was pulled away.
4. The trainer requested the arm extension behavior be performed several times before injecting with a sharp needle. In the beginning, the animals reacted to this with a surprised squeak, but usually remained seated and were willing to continue extending their arms.

For the first male, 50 training sessions distributed over 15 weeks passed before he cooperated during injection. This was a time investment of approximately five hours of actual training. The training progressed more rapidly with two other subjects. One of them reached the goal after 90 minutes, the other after four hours of training distributed over 18 sessions and 43 sessions, respectively.

The three trained males did not show signs of fear or resistance during the injection procedure, and they all cooperated not only with the trainer but also with other personnel.

4.4.2.2. Training to Cooperate during Sample Collection from Vascular Access Ports

Friscino et al. (2003) surgically instrumented three female and nine male **rhesus macaques** with biliary and venous catheters that could be accessed in a pouch located on the back of the subjects' jackets. The animals were then trained—using an unspecified positive reinforcement protocol—in their home cages to present the pouch and to remain stationary while the catheters were accessed. Three to four training sessions spread over a two-week period were required to achieve cooperation. The successful training precluded the need to subject the animals to enforced manual restraint or chair-restraint during sample collection.

4.4.2.3. Training to Cooperate During Saliva Collection

Tiefenbacher et al. (2003) presented nine individually housed adult male **squirrel monkeys** in their home cages with a thin, 10 cm long PVC pipe to which a braided cotton dental rope was attached on one end, and a plastic-coated cable—for retrieving the device—on the other end. The dental rope was flavored by soaking it in a solution of one part Kool-Aid®, one part sugar, and three parts water; it was then baked to dryness.

Seven of the nine monkeys readily acquired the task of chewing on the cotton rope for at least 30 seconds, after which the device was retrieved and the subjects were rewarded with a food treat. The saliva obtained in this manner was sufficient to permit cortisol analysis by RIA (radio-immuno assay). Two monkeys required the addition of peanut butter and/or jelly to the dental rope to elicit sufficient chewing; only one monkey refused to cooperate in this saliva collection technique. Repeated saliva samples could be obtained reliably from the other eight animals.

This technique may also lend itself to the non-invasive assessment of other hormones and compounds in saliva. It was originally developed for rhesus macaque infants (Boyce et al., 1995) and adapted to adult **rhesus macaques** by Lutz et al. (2000) who found that 21 of 23 subjects cooperated, but only 16 (76 percent) produced saliva samples that were sufficiently large (0.4 ml) to allow cortisol analysis.

Cross et al. (2004) found in four adult male and five adult female **marmosets** (*Callithrix* sp.) that adequate saliva samples for RIA assessment of cortisol can be obtained reliably, without any extra training, by presenting the animals nine times for a cumulative total of approximately five minutes a cotton-wool bud coated with a thin layer of fresh banana. The animals spontaneously lick and chew on the bud. Many substances such as fruit-drink crystals, gum arabicum, honey, sugar water and crushed mealworm were tried as an alternative to banana to tempt the marmosets to lick and chew the cotton-wool buts, but banana was found to be the only substance that reliably encouraged chewing.

4.4.2.4. Training to Cooperate During Semen Collection

Brown (1998) and Brown and Loskutoff (1998) document and describe how they trained three adult male **gorillas** *(Gorilla gorilla gorilla)*, living together as a bachelor group, to cooperate during semen collection rather than subjecting the animals to electro-ejaculation under general anesthesia:

> *The gorillas were not forced into the training area nor did they have to cooperate with the trainer. The training area was an off-display holding cage with a 7.5 x 15-cm opening covered by a solid plate steel sliding door at ground level. The training was based on shaping behavioral responses with positive reinforcement using verbal and food rewards.*
>
> *1. The first behavior introduced was "Station." The trainer said "Station," the gorilla approached and took a treat with his lips. As training progressed, rewards were withheld until the gorilla approached, sat down directly in front of and facing the trainer, and accepted a treat in response to the "Station" prompt. All the remaining behaviors were taught with the gorilla in the "Station" position.*
>
> *2. The verbal prompt "Target" was used to associate an object with a desired response. The prompt was given while touching a ping pong paddle to the gorilla's fingers opportunistically, when the gorilla placed his hands on the wire mesh. Soon, the gorilla touched the paddle as a response to the "Target" prompt.*
>
> *3. The verbal prompt "Hold" was added to the "Target" behavior. When the gorilla touched the paddle through the mesh, the trainer said "Hold," while lifting the paddle off the mesh and moving it out of sight. The "Hold" behavior was shaped so the gorilla remained in the "Target" position until the trainer gave the bridge "Okay," while administering a reward.*
>
> *4. The cue "Knee" was shaped with the gorilla in the "Hold" position. The trainer passed a 70 cm long, 2.5 cm-diameter PVC pipe through the mesh and touched the knee when the verbal prompt was given. As training progressed, the gorilla moved the requested body part to the pipe. Eventually, he responded by moving the knee to the finger tips of the trainer's hand. This completed the shaping of the "Knee."*
>
> *5. Before semen collection was attempted each gorilla performed reliably the following "set up" procedure: "Station," "Target," "Hold," "Knee," and "Hold."*
>
> *6. Originally, an artificial vagina, constructed using a PVC pipe, was placed on the gorilla's penis. After numerous attempts, however, it was not tolerated by the gorillas nor did it stimulate ejaculation. It was decided that the trainer needed to reach through the small door with the left hand*

and stimulate the genital area directly. If the animal broke the "Hold" position, the trainer immediately withdrew, closed the door, and repeated the "set up" procedure. Eventually, with continuous administration of treats and repetition of the "Hold" prompt, the gorilla allowed penile massage periods long enough to result in ejaculation. As soon as the ejaculate was collected, the trainer's hand was withdrawn, the sliding door was shut, and verbal praise and treats were given to the gorilla.

7. One of the goals of the training program was to provide a positive experience for the gorillas. Nevertheless, when they were unruly or uncooperative, two types of discipline were used. The first, the most common, was verbal. Verbal discipline included stating the gorilla's name and saying "No" in a low, strong voice. Never was the verbal reprimand shouted. The other type of discipline used was "time out," given when the verbal reprimand failed twice. In these instances, training ceased, and all personnel exited and remained silent and out of visual contact with the animal for 1-3 minutes. Upon return, training resumed normally. The gorilla usually cooperated with the trainer after having a "time out," but if he did not, the trainer continued to give prompts until the gorilla performed a requested behavior. The gorilla was immediately rewarded, and the training session ended on a positive note.

Training sessions were 10 to 20 minutes long on three days per week. The first semen samples of the three gorillas were obtained five, 12 and 14 months after starting the initial training.

4.4.2.5. Training to Cooperate During Blood Pressure Measurement

Mitchell et al. (1980) trained three single-caged adult male and one adult female **baboons** to voluntarily submit to self-initiated blood pressure measurement in their home cages. Traditionally, blood pressure measurement involves considerable distress for the animals who first have to undergo surgery for arterial catheterization and are then chair-restrained against their will during data collection.

A cage-mounted oscillometric instrument with adjustable cuff assembly and banana-flavored pellet rewards was used for the training. Subjects were rewarded for extending their forearms into the cuff and depressing a lever to initiate measurement and maintain arm position throughout the blood pressure measurement sequence. Releasing the lever or withdrawing the arm too early caused immediate venting of cuff pressure and withholding of the reward.

Initially, the animals' tasks was simplified by mounting the lever directly against the front panel of the cage. This caused the lever to protrude slightly into the cage, where the subjects could reach it with minimal arm extension. In addition, the lever depression time required to earn one pellet was set at about 0.1 second and was then gradually increased to about 35 seconds. Only then was the cuff assembly installed and were the subjects rewarded for fully extending their forearms in the cuff and depressing the lever during a normal blood pressure determination.

All four baboons were trained successfully to cooperate during this procedure. The number of training sessions, which averaged 60 minutes each, ranged from 35 to 51, with a mean of 43.

Turkkan et al. (1989) and Turkkan (1990) trained 10 adult male baboons to cooperate during blood pressure measurements in their home cages. Training occurred before the daily pellet ration was distributed to ensure that food rewards during training were salient reinforcements. The following training protocol was applied:

After an arm shelf with a 12-cm post at one end is attached to the subject's cage, the animal is rewarded with a food pellet for the following actions in progression:

1. *Extending an arm onto the shelf.*
2. *Extending the left arm as far as the post.*
3. *Touching the post.*
4. *Grasping the post with the left hand.*
5. *Holding the post for increasing durations.*
6. *Allowing the arm that is holding the post to be touched.*
7. *Allowing the arm to be stroked with the blood pressure cuff.*
8. *Allowing the cuff to be placed briefly around the arm. At this stage, the cuff is opened and closed repeatedly so that the animal will habituate to the sound of the Velcro fastening and unfastening.*
9. *At each step, food rewards are given freely.*
10. *With the cuff in place, allowing the stethoscope to touch the extended arm (***Figure 61a***).*
11. *With the cuff, stethoscope and aneroid manometer in place, the trainer slowly inflates the cuff while delivering frequent food pellets (***Figure 61b***). It is important to keep the training session short so that aversion to the cuff inflation does not have time to develop. Most animals begin to pull on the blood pressure apparatus at this stage, and the trainer needs a quick hand to rescue all the paraphernalia before the animal can pull them into his cage. Also at this stage, training is facilitated by switching from food pellets to fresh fruit chunks, or applesauce dispensed to a food nozzle by means of an infusion pump. The applesauce has the added advantage of providing immediate termination of a continuous stream of reinforcement when inappropriate behavior such as arm withdrawal occurs.*
12. *The rate and degree of cuff inflation is progressively increased over successive sessions, with termination of applesauce reinforcement for arm withdrawal, which occurred frequently at this stage.*

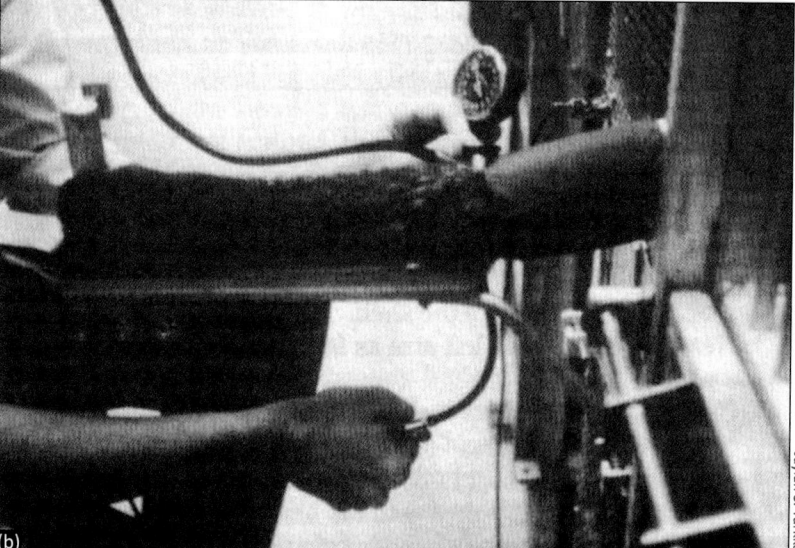

Figure 61a,b. Adult baboon *Jim* cooperates during manual auscultatory blood pressure measurement; note that there is no squeeze-back forcing the animal to sit still at the front of the cage. *Jim* voluntarily extends his arm and holds the post at the end of the shelf; the cuff is placed on the arm, the stethoscope on the brachial artery (a), and the cuff is inflated (b).

13. *An episode of uncooperativeness must never be allowed to end a training session, because then the animal quickly learns to avoid discomfort by acting aggressively. When aggressive acts such as scratching the trainer occur, the reinforcement is withheld, and the training resumed after a few minutes.*
14. *Once the baboon accepts full cuff inflation, the cuff is deflated slowly. The animal is rewarded for sitting through a period of non-reward while the trainer attends to blood pressure measurement.*
15. *After completion of the final measurement, the baboon is rewarded with fresh fruit.*

The duration of training until the first systolic and diastolic blood pressure measurements were obtained was an average 12 weeks (range 2 to 36 weeks). Systolic and diastolic blood pressure measurements of a trained baboon required approximately five minutes, which included set-up of the shelf and food reward delivery system.

4.4.2.6. Training to Cooperate During Oral Drug Administration

Oral drugs are traditionally delivered via gavage, which is one of the most distressing procedures to which nonhuman primates are subjected.

Turkkan et al. (1989) habituated 11 adult **baboons** of unspecified gender to voluntarily drink a bitter-tasting solution of quinine which could mask the taste of various test drugs. Initially, the subjects were offered 100 ml of an orange-flavored juice that they all drank avidly. Over daily sessions, increasing amounts of quinine sulfate were added to this orange drink until a concentration of 0.325 mg/ml was reached. It was then possible to add test drugs and to obtain complete dose-effect curves with a number of benzodiazepines, barbiturates and other sedative/anxiolytic drugs. Unfortunately, the authors do not indicate how much time was invested to successfully habituate the baboons to drink the quinine solution.

Baumans et al. (2007) quote a report on **vervet monkeys** (*Cercopithecus aethiops*) who voluntarily swallow drugs when these are mixed with the animals' regular diet, consisting of pre-cooked maize fortified with vitamins and minerals. The dry ingredients are blended with water and form a stiff putty-like paste, which is an ideal vehicle for mixing in test substances. If the flavor needs to be masked, there are a variety of possibilities, such as honey and syrup, depending on what the protocol permits:

We usually administer the compound in about a third of the morning feed. The bulk of the food is offered after this portion has been consumed. Some substances we even mix into the entire bulk of the morning feed. Keeping the compound too long in cheek pouches or spitting it out has never been a problem. We have used this simple oral administration technique for pharmacokinetic studies very successfully. Over a time period of 20 years, we have not had to deal with any substance that we could not feed to the vervets, including bitter herbal mixtures in fairly high concentrations.

4.4.2.7. Training to Cooperate During Topical Drug Application

Reinhardt and Cowley (1990) trained adult **stump-tailed macaques** to actively cooperate during drug application on their foreheads in the home cages. The animals were used to being removed from their cages and subjected to enforced mechanical restraint during this procedure in a treatment area.

There were 17 males and three females living in 10 compatible pairs in double-cages equipped with sturdy, replaceable plastic plates that fitted into the cage door openings. Each plate had a face-shaped hole fitting the head of an animal and two smaller circular holes fitting the forearms. The arrangement of the holes was such that an animal could reach out for raisins and eat them while presenting his or her forehead (**Figure 62**). For the treatment, the pairs were temporarily separated by means of a cage divider so that one partner could be treated without the other interfering.

The animals required one to 14 training sessions, each lasting one to five minutes, to present their foreheads and allow topical drug application while retrieving raisins from the handler's hand.

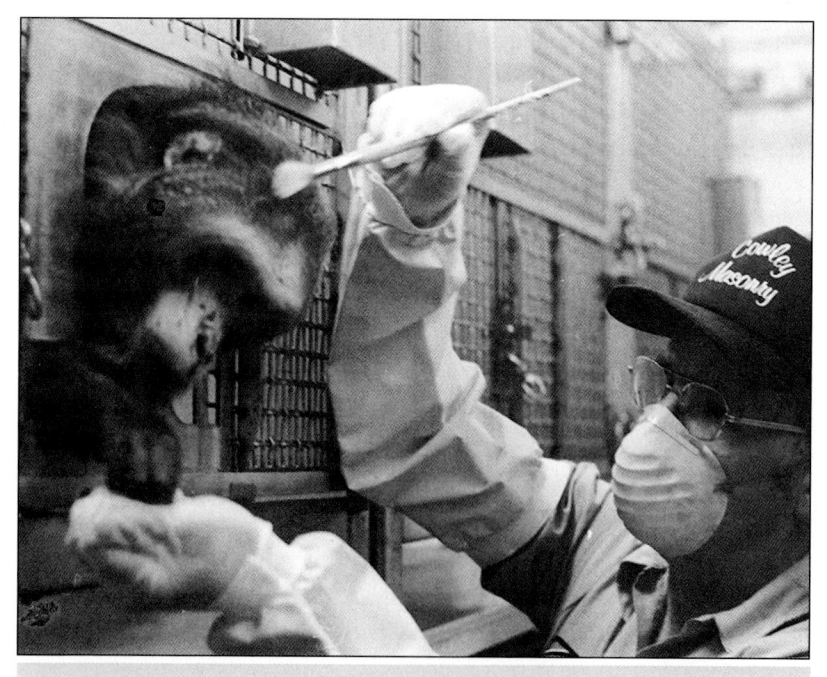

Figure 62. Stump-tailed macaque *Stan* cooperates during topical drug application in his home cage.

4.4.2.8. Pole-and-Collar-and-Chair Training

It has been repeatedly stated that monkeys can be trained to voluntarily cooperate in their home cages to have a pole or leash attached to a collar and allow themselves to be subsequently guided to and securely placed in restraint chairs (Barrow et al., 1966; Nahon, 1968; Anderson and Houghton, 1983; Schmidt et al., 1989b; McCully and Godwin, 1992; Klein and Murray, 1995; Marks et al., 2000; Sauceda and Schmidt, 2000; Scott et al., 2002; Down et al., 2005). This claim is supported with data in only one case.

Skoumbourdis EK (2008) has trained adult and juvenile rhesus macaques and adult long-tailed macaques to cooperate during the capture with the pole and the transfer to and placement in the restraint chair:

All the monkeys I have pole/collar/chair trained have gone through an initial phase of resistance both when the pole was being attached to the collar, and when they were first put into the chair, but for the most part they finally did settle down and cooperate. All it takes is patience and gentle determination on the part of the trainer.

I always collar my animals at least a week or two before the first training session so they get used to wearing the collar. If they're not comfortable with the collar, it really sets the training back because they will spend most of their time pulling at the collar and scratching at their neck.

To start the training, I first make sure that the trainee is comfortable enough with me that he/she is willing to take treats from my hand. I subsequently include the pole, offering treats with one hand, while holding the pole close to the cage in the other. The animals readily get used to this little ceremony and soon seem to ignore the pole, but focus more on the treats.

The poles come with that handy little clip that opens and closes for collar attachment. The clip is a great place to hook treats, which the monkey has to retrieve directly from the "dreaded pole." I like to stuff a marshmallow tightly into the clip. This makes it a little harder for the animal to get the treat, and extends the time the animal is in contact with the pole. Once the monkey retrieves treats consistently, without signs of apprehension or fear, I start moving the un-baited pole very carefully in the cage, and finally, also touch the animal with it. In subsequent sessions, I gently tap the collar with the pole. When the training session is over, I hang the pole outside on the front of the cage so that the animal gets more and more acquainted with it. Needless to say that I always distribute extra rewards—jackpot if it's deserved!—before I leave the room.

I have trained animals living in both, cages equipped with squeeze-backs and those without. If the animal's cage has a squeeze-back, I use it only with the tougher customers. But, generally, I try to avoid using it so that the trainee is always in control of the situation. I believe this greatly helps the animals to stay relaxed, continue to trust me, and learn quickly what is expected from them in each training session. I also consistently reward cooperation with a treat and verbal praise. If the animal doesn't cooperate, patience from my part replaces the reward. I feel that this strategy helps to create a tension-free ambience for the monkey and for the trainer.

The first few times the pole is actually attached to the collar can be quite dramatic. The trainees usually "freak out" the moment they realize what is happening to them. However, there is no reason for panic. I simply leave the pole attached, maintain a firm grip, and talk reassuringly to the animal who will gradually calm down, stop squirming, and remain quiet long enough so that I carefully unhook and remove the pole. This interaction is always followed by a generous treat reward which, in my experience, is never refused.

During the next sessions, I get the trainee to sit still with the pole attached to the collar for progressively extended periods of time, until he/she "forgets" about the pole and takes treats from me. I repeat this step several times. Some animals adjust better to this situation than others, but they all end up remaining reasonably still with the pole attached to the collar.

Coaxing the poled monkey to come out of the cage is always a big challenge. After all, the familiar home cage is a relatively safe haven for these animals. With patience, and many reassuring words, the trainee does finally stop resisting, follows the pull of the pole, and comes out of the cage. Should the animal begin to thrash about once outside of the cage, I take the pole and carefully, but firmly, push the animal's head to the floor. To be clear, I do not throw him/her down but rather use the pole to turn the collar up towards the animal's head and then apply some forward and downward pressure in a determined manner. The monkey is now fixed and can get his/her bearings while remaining safe from causing himself or herself any serious harm. I have noticed over and over again that you can help the animal to calm down when you speak to him/her reassuringly with a gentle whisper-like voice. When the animal has settled down, I carefully start to walk him or her again; I will drop a few treats on the floor for the animal to pick up as he/she moves along the floor. After a few sessions, most trainees will feel confident enough to walk, rather than struggle, on the pole. If a monkey continues to resist after two or three sessions, I'll call in reinforcements. Most collars have two sides where a pole can be attached. By adding a second pole, directed by a second person, the animal is easier to guide in a forward motion.

I've found that it takes about one week of training until a monkey will cooperate and walk on the pole in a reasonably calm manner and pick up treats from the floor as a reward for good behavior. My goal is to get the trainees to walk, because after they come out of their cages they have a lot of pent-up energy that they like to release; especially the younger animals. I treat this solely as a reward for good behavior. If the poled animals walk calmly, I let them do so for a few minutes, but if they start playing "super man," I pull them straight back into their cages. If you don't have enough space, or the racks are enticingly close for climbing and rattling, or if you are a little new at this and do not have a second person around who can help you control the monkey if need arises, the pole walking isn't a good idea.

Now, onto the chair:

1. *Push the chair up against a wall, with the entrance facing out, and put all the brakes on. This keeps the chair stable and makes it impossible for the monkey to walk straight through—a situation that isn't any fun when you're on the other end of the pole!*
2. *Allow the monkey to explore the chair, touch it, climb it, walk around it, and perhaps retrieve a treat or two that you have placed somewhere on the chair.*
3. *After a day or so, coax the monkey into the sitting position in the chair. Do this by gently lifting the animal's neck into position and get the collar into place. If another person, who is also on very good terms with the trainee, can help you, the situation becomes less of a challenge, especially when you are dealing with a strong and extremely stubborn monkey. Once you have your monkey in place, let him/her adjust for a few minutes. Don't forget the treats! Some animals will be initially restless and try to push your hand away; but with gentle patience they will settle down and finally accept the food reward.*
4. *Gradually extend the time the trainee remains in the chair over the next few days. Always be sure to remain close by to serve as a comforting social support. Should the animal show any signs of discomfort, try giving him/her further treat rewards. If he or she continues to be restless, abort the training session; you do not want the animal to relate the chair with discomfort and/or distress.*

I have found that each "big step" involves an initial struggle, but I have also found that with consistency and patience, the animals learn quite quickly what I expect them to do. I have had several animals who were fully trained and just came up to the front of the cage without being squeezed. They actually presented their necks so that their collar loop was exposed for me to attach the hook of the pole. All of these monkeys struggled a great deal when I first started working with them. It is amazing how these animals gradually relax into the training sessions and finally start working with you, rather than against you.

Trust in the trainer is the ultimate key for success. Nonhuman primates are intelligent; when they are free of apprehension or fear, they quickly figure out that it is much easier and even rewarding for them to cooperate with you rather than resist. A successfully trained monkey will have developed so much trust in you that he/she will never fight against you when you pole and chair him/her.

When I train animals, I work with them once or twice daily, five days a week—with additional weekend sessions if needed—until the goal of the training has been achieved. I have found that if I don't work with them on a consistent schedule, they tend to get "rusty" rather quickly. The faster you can get them over the initial struggling, the easier the whole training sequence. If you try to pole a monkey who vigorously resists on a Monday, and decide to wait and try again on Friday, chances are that the struggle will be the same, if not worse. However, if you are persistent and repeat the training step over and over again every day, you will definitely notice progress by the end of the week. I imagine that without consistency and patience, the training would be a rather frustrating experience, both for the trainer and for the trainee.

To pole-collar-chair train a monkey can be a very rewarding process that is not necessarily time-consuming. I have successfully trained 19 animals:
    two adult female rhesus,
    four adult male rhesus,
    five juvenile male rhesus,
    four adult female cynomolgus, and
    four adult male cynomolgus.

My quickest subject took just five days of training to reliably cooperate (I should mention that he was two years old and an angel!), while other animals have taken me well over a month to get going—especially older rhesus who can be very stubborn and hard to food-motivate. Also, I have had some animals who were just never meant to be put in a chair. This is a reality that both you and the investigators must acknowledge. You cannot force a monkey to cooperate and be relaxed in the chair. It's impossible. Sure, you can try, but you're not going to win.

4.4.2.9. Training to Cooperate for Weighing

McKinley et al. (2003) trained six heterosexual pairs of **marmosets** to cooperate for weighing in the animals' home cages rather than being caught by gloved hands and transferred to a small cage to be weighed:

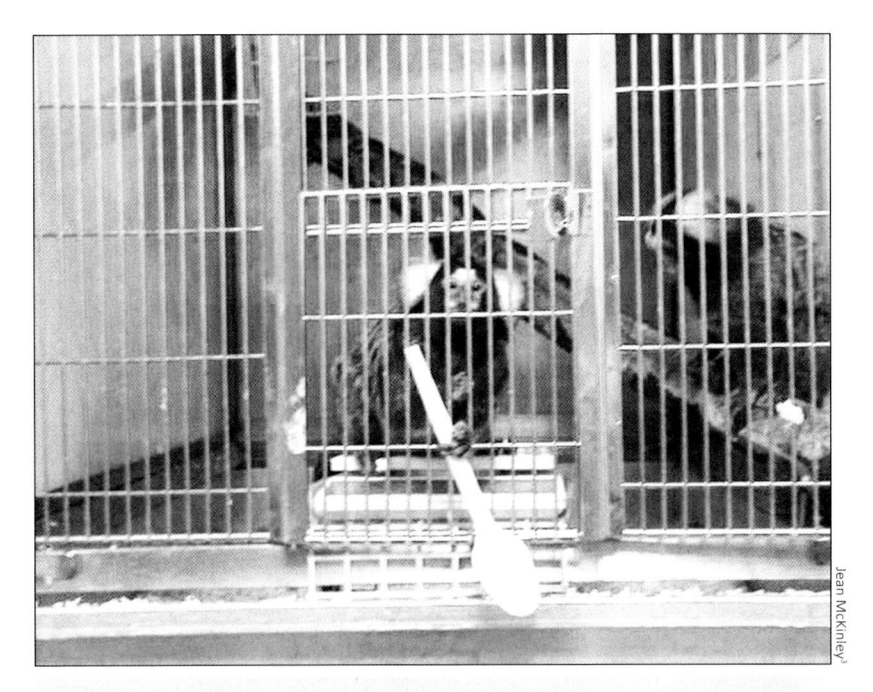

Figure 63. This target-trained female marmoset sits on scales in the familiar home cage while her body weight is recorded; her partner waits until the target is presented to him.

1. *The target* (a plastic spoon) *was held at the front of the cage with the food reward* (marshmallow, cornflakes or chopped dates) *held behind it. Males were offered a black target placed on the left-hand side and females a white target placed on the right. A reward was given when the correct target was touched. Incorrect responses were ignored.*
2. *The target was presented without the reward held behind it.* The animals were rewarded when they touched the target.
3. *The time the target had to be held before the reward was given was gradually increased.*
4. *Scales for weighing were placed in the cage and the target held in front of them. The marmoset was rewarded for climbing onto the scales and holding the target* while her or his body weight was recorded (**Figure 63**).

The cumulative time per animal to achieve the goal of the training ranged from 20 to 120 minutes with a mean of 64 minutes. The time investment for successful training did not differ between females and males.

[3]reproduced with permission from *Journal of Applied Animal Welfare Science* 6(3), 209-220, 2003.

Figure 64a,b. Paired rhesus macaques entering a transfer box—one at a time—on vocal commands (Reinhardt, 1992c). Note that the animals are not forced with a squeeze-back or a stick to leave their cage

4.4.2.10. Training to Cooperate for Capture

Traditionally, mechanical force (movable squeeze-back), threats (display of net) and vocal intimidation are used to overcome the reluctance of primates to leave their familiar home cages while these are sanitized or for routine procedures such as weighing. It has been reported that monkeys can be trained to voluntarily exit into transfer boxes (**Figure 64a,b**; Clarke et al., 1988; Heath, 1989; Sainsbury et al., 1990; Reinhardt, 1992c; Erkert, 1999; White et al., 2000; Coke et al., 2007); detailed training protocols have yet to be published.

5. DISCUSSION

Distress in laboratory animals is usually unnecessary (Institute for Laboratory Animal Research, 1992, p 85).

The literature makes it clear that the distress resulting from involuntary permanent confinement in a standard barren cage can be alleviated by providing the imprisoned primate with:
• compatible companionship,
• foraging opportunities, and
• access to the "safe" vertical dimension.

5.1. Compatible Companionship

Group-housing would be the most species-appropriate refinement alternative to single-housing. Safe procedures of transferring single-caged individuals to compatible group-housing arrangements have been documented for pig-tailed macaques, long-tailed macaques and chimpanzees. There is good reason to believe that other species, such as baboons, stump-tailed macaques, squirrel monkeys, capuchin monkeys and common marmosets, can also be transferred from single-housing to compatible group-housing arrangements if basic ethological principles are applied. Attempts with rhesus macaques have so far been discouraging. This species, as probably all other non-human primates species, can readily be transferred from single- to social-housing conditions by carefully pairing adult individuals with same-sex companions (to avoid uncontrolled breeding) or with naturally weaned infant companions. Compatible pair-housing has the advantage over group-housing that individual subjects are readily accessible and that it does not interfere with common research protocols.

There is a professional consensus that:

a compatible conspecific probably provides more appropriate stimulation to a captive primate than any other potential environmental enrichment factor (International Primatological Society, 1993, p 11).

National and international regulations and guidelines have incorporated this assumption in their stipulations and recommendations:

1. *Any primate housed alone will probably* **suffer** [emphasis added] *from social deprivation, the stress from which may distort processes, both physiological and behavioural* (Canadian Council on Animal Care, 1984, p 165).

2. *Social interaction is paramount for well-being. Social deprivation in all its forms must be avoided. Isolation can only be justified for* **short** [emphasis added] *periods during the experimental procedure or during essential veterinary treatment* (National Health and Medical Research Council, 1997, p 3 & 5).
3. *Primates are very social animals. Physical contact, such as grooming, and non-contact communication through visual, auditory, and olfactory signals are vital elements of their lives. Providing animals with a satisfactory social interaction helps to buffer against the effects of stress, reduce behavioral abnormalities, increase opportunities for exercise and helps to develop physical and social competence* (Primate Research Institute, 2003, Chapter IV).
4. *Pair or group housing* **must** *be considered* **the norm** [emphasis added]. *For experimental animals, where housing in groups is not possible, keeping them in compatible pairs is a viable alternative social arrangement. Single caging should only be allowed where there is an approved protocol justification on* **veterinary** *or* **welfare** [emphasis added] *grounds* (International Primatological Society, 2007, p 11).
5. *Primates should be socially housed as compatible pairs or groups. They should not be singly housed unless there is* **exceptional** [emphasis added] *scientific or veterinary justification* (Medical Research Council, 2004, p 6-8).
6. *The remarkable sociality of the primate order in general is the most relevant characteristic for their* **humane** [emphasis added] *housing* (US Department of Agriculture, 1999, p 17).
7. *The environmental enhancement plan* **must** [emphasis added] *include specific provisions to address the social needs of nonhuman primates* (US Department of Agriculture, 1995, §3.81(a)).
8. *Single housing should only occur if there is justification on* **veterinary** *or* **welfare** [emphasis added] *grounds. Single housing on experimental grounds should be determined in consultation with the animal technician and with the competent person charged with advisory duties in relation to the well-being of the animals* (Council of Europe, 2006, p 14).

Despite the significant importance of housing primates in a social setting rather than alone, social caging has yet to become implemented as a standard refinement practice:

- *Single or individual caging systems are the basic or staple housing used for primates. Almost all 'hard' scientific data have been acquired from singly caged primates* (Rosenberg and Kesel, 1994, p 459 & 460).
- *The common practice of housing rhesus monkeys singly calls for special attention* (National Research Council, 1998, p 99).

Two independent surveys of primate facilities located in the United States revealed that the percentage of indoor caged macaques housed socially did *not* increase over a time period of nine years (**Table 2**). Both in 1994 and 2003, only about one third of the animals lived with one or several partners, while two thirds were living alone (Baker et al., 2007).

	1994	2003
Rhesus macaques	56 percent	48 percent
Long-tailed macaques	16 percent	33 percent
Pig-tailed macaques	23 percent	15 percent
Mean	**32 percent**	**34 percent**

Table 2. Percentage of indoor caged macaques housed in US facilities with one or several companions in a 1994-survey (Reinhardt, 1994) and in a 2003-survey (Baker et al., 2007).

Some primatologists have taken the side of the single-caging practice, probably because any changes to this traditional housing practice could invalidate the precious historic database (Dean, 1999) and upgrading the standard caging system would require extra funds (Crockett, 1993; Crockett and Bowden, 1994).

The following arguments have been brought forth against the transfer of single-caged primates—especially rhesus macaques—to social-housing arrangements:

1. *The rhesus monkey is extremely nervous and energetic and is difficult to house.* ***Unquestionably*** [emphasis added], *animals involved in experiments should be housed in individual cages* (Gisler et al., 1960, p 760).
2. ***Any** [emphasis added] plan to increase social interaction also increases the risk of injury and death. Unless they have grown up in the same social group, primates are not likely to tolerate each other when placed together as adults. Besides the risk of trauma, there are other disadvantages to allowing increased social interaction. Contact between animals may lead to greater transmission of infectious diseases* (Line, 1987, p 858).
3. *Especially when new pairs are formed and dominance relationships are being established, there is a strong likelihood that the **veterinarian will be kept quite busy suturing wounds*** [emphasis added] (Coe, 1991, p 79).
4. *When adult rhesus monkeys are first paired there are **always*** [emphasis added] *injuries incurred* (Rosenberg and Kesel, 1994, p. 470).
5. *The possible behavioral advantages of pair housing may be offset by the increased potential of contagious diseases, for wounding, and for undernourishment in the less dominant partner* (Novak and Suomi, 1988, p 769).
6. *Pairing is not uniformly beneficial, however. The animals usually form dominance relationships, and the subordinate partner may be subject to behavioral depression or distress* (Line et al., 1989, p 105).
7. *Social pairing **is** [emphasis added] associated with high health risks to monkeys* (Morgan et al., 1998, p 168).
8. *Long-term housing with the same partner may sometimes lead to boredom, as expressed by a decline in social interaction and an increase in general passivity* (Novak and Suomi, 1988, p 770).

The reviewed published data make it quite clear that nonhuman primates—including rhesus monkeys—can readily be transferred from single- to pair-housing, and some species to group-housing settings if basic ethological principles are applied to minimize the risk of injurious aggression related to the establishment of dominance-subordinance relationships.

Published data also indicate that the health risks tend to decrease rather than increase when single-caged animals are transferred to compatible pair-housing arrangements. There is not one published record demonstrating that subordinate partners of compatible pairs suffer from undernourishment; this is probably due to the fact that food sharing is one criteria of partner compatibility. There is also no published case showing that long-term pair-housing with the same partner leads to boredom, with the two companions showing a decline in their motivation to interact with each other.

Being separated from each other during post-operative recovery, food-intake, metabolic and neurophysiological studies is likely to distress paired companions. The published literature offers practical guidance on how partner separation can be avoided during common research protocols without jeopardizing the safety of the animals and the scientific integrity of the study.

The transfer to compatible social-housing provides previously single-caged primates not only with a living environment that can cure them from the behavioral pathology of self-injurious biting and help them cope with potentially distressing situations, but it also enhances their general well-being by allowing them to be what they truly are: social rather than solitary animals. Living with one or several conspecifics makes it possible for the caged primate to actively express his or her biologically inherent need to engage in social behaviors.

5.2. Foraging Opportunities

The reviewed literature offers numerous options making it possible for caged primates to get more involved in food searching, food retrieving, and food processing activities, thereby allowing them, at least partially, to satisfy their biological urge to forage. The most practical, least expensive, yet effective way of feeding enrichment is the presentation of the daily food ration in such a way that the animals can work for it.

The importance of foraging opportunities for the well-being of caged nonhuman primates is underscored and clearly addressed by some professional guidelines and legal rules, while others do acknowledge foraging behavior but fail to recommend that it should be actively encouraged in captive animals.

- The International Primatological Society (1993, p 9-10) recommends in its *Codes of Practice* that:
  Opportunities should be provided for primates to express most normal behavior patterns. *Opportunities for increased foraging are ranked as the first, most important ones of particular benefit. Foraging time can be increased by providing some of the animal's food in such a way as to make its delivery or discovery unpredictable. As animals like to work for their food, increasing processing time, increasing foraging, or providing puzzle feeders or other feeding devices is encouraged* (International Primatological Society, 2007, p 16).
- The Medical Research Council (2004, p 9) states in its *Best Practice in the Accommodation and Care of Primates used in Scientific Research* that:

> *Foraging enhances welfare and minimizes the expression of abnormal behaviors. Therefore, all primates should be given the opportunity to forage daily, by scattering food in litter or substrate on the floor, or in a tray, and by using devices that encourage foraging activity (e.g., puzzle feeders).* The Medical Research Council *will require justification for the use of scientific procedures that restrict the opportunity to forage.*

- The Council of Europe (2006, p 48) stipulates in its *Appendix A of the European Convention for the Protection of Vertebrate Animals Used for Experimental and Other Scientific Purposes (ETS No. 123)* that:

  Presentation and content of the diet should be varied to provide interest and environmental enrichment. Scattered food will encourage foraging, or where this is difficult, food should be provided which requires manipulation, such as whole fruits or vegetables, or puzzle-feeders can be provided.

- The US Department of Agriculture (1995, §3.81(b)) lists in its Animal Welfare Regulations for nonhuman primates:

  varied food items, using foraging or task-oriented feeding methods as examples of environmental enrichment,

  but falls short to stipulate that such methods should be an integral part of the environmental enhancement plan.

- The National Research Council (1996, 1998) does not offer clear guidance and fails to recommend the provision of foraging possibilities for nonhuman primates:

  1. The National Research Council's *Guide for the Care and Use of Laboratory Animals* (1996, p 40) simply notes that:

     In some species (such as nonhuman primates) and on **some** *[emphasis added] occasions, varying nutritionally balanced diets and providing "treats," including fresh vegetables,* **can** *[emphasis added] be appropriate and improve well-being.*

  2. The National Research Council's book, *The Psychological Well-Being of Nonhuman Primates* (1998, p 39), briefly mentions that:

     Feeding **can** *[emphasis added] be used to provide positive behavioral stimulation as a means of enhancing primate well-being.*

5.3. Access to the Vertical Dimension

There is a professional and regulatory consensus that caged nonhuman primates need to have access to high structures in order to feel relatively safe:

1. *Under natural conditions, many primates spend much of their lives above ground and escape upward to avoid terrestrial threats. Therefore, these animals might perceive the presence of humans above them as particularly threatening* (National Research Council, 1998, p 118).

2. *The vertical dimension of the cage is of importance and cages where the monkey is able to perch **above** [emphasis added] human eye level are recommended* (International Primatological Society, 1993, p 11).
3. *Cages should be designed and constructed so that the space [is] enough to allow for an appropriate rest structure* (Primate Research Institute, 2003, Chapter VI). *Perches and three-dimensional structures should be arranged to make as much use of the available space as is possible* (Primate Research Institute, 2003, Chapter IV).
4. *The volume and height of the cage are particularly important for macaques and marmosets, which flee upwards when alarmed. Their cages should be floor-to-ceiling high whenever possible, allowing the animals to move up to heights where they feel secure. Double-tiered cages should not be used since they restrict the amount of vertical space available to the animals* (Medical Research Council, 2004, p 7). *A two-tiered system is not recommended as these cages are usually too small. The lower tiers do not allow primates to engage in their vertical flight response, are often darker, and animals in the lower cages tend to receive less attention from attending personnel* (International Primatological Society, 2007, p 12).
5. *The flight reaction of non-human primates from terrestrial predators is vertical, rather than horizontal; even the least arboreal species seek refuge in trees or on cliff faces. As a result, enclosure height should be adequate to allow the animal*

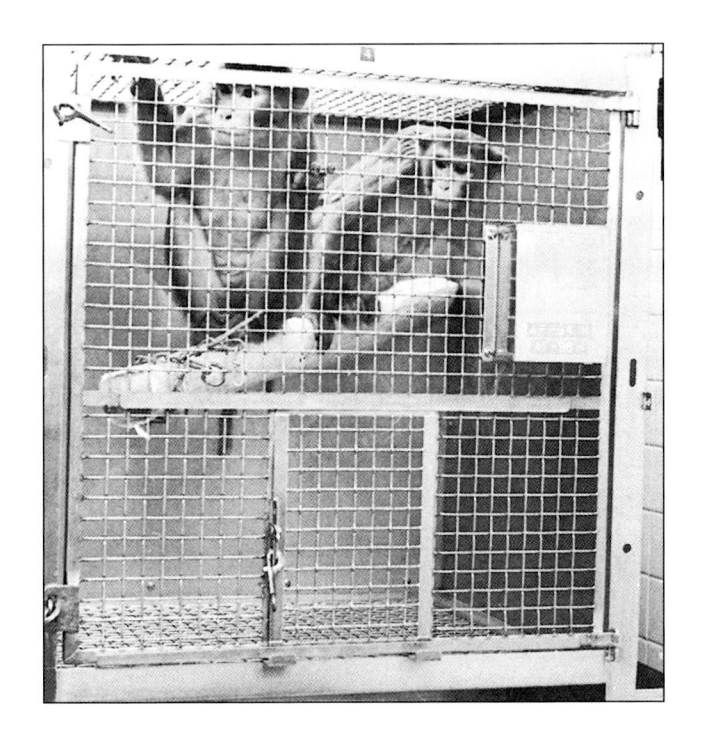

*to perch at a sufficiently high level for it to feel secure. The minimum enclosure height for caged marmosets and tamarins is 1.5 m; the minimum enclosure height for caged squirrel monkeys, macaques, vervets and baboons is **1.8 m** [emphasis added]. It is essential that the animals should be able to utilize as much of the volume as possible because, being arboreal, they occupy a three-dimensional space. To make this possible, perches and climbing structures should be provided* (Council of Europe, 2006, p 42,52,54).

Access to the vertical dimension addresses the caged monkey's biological urge to retreat to and rest in the relatively safe arboreal dimension of the living quarters. Animal welfare regulations downplay the importance of elevated resting surfaces, such as perches, when they merely list these as optional examples of environmental enrichments (US Department of Agriculture, 1995, §3.81(b)).

A high perch does not really "enrich" the environment of a caged primate but it is a necessity for the animal and, hence, should be a mandatory standard furniture of every cage in which nonhuman primates are kept. The reviewed literature attests that high perches can easily be installed both in standard and squeeze-back cages and that the animals do make consistent use of them.

5.4. Positive Reinforcement Training

It is obvious that a monkey or ape is distressed when he or she is removed from the familiar home cage, forcefully restrained and then subjected to a life-threatening procedure such as injection or venipuncture. It is also obvious that a monkey or ape is less distressed or not distressed at all when he or she has been trained to cooperate, rather than resist during handling procedures. Professional guidelines and regulatory stipulations take this circumstance into consideration:

1. *Procedures that reduce reliance on forced restraint are less stressful for animals and staff, safer for both, and generally more efficient* (National Research Council, 1998, p 46).
2. *Restraint procedures should only be invoked after all other less stressful procedures have been rejected as alternatives* (Canadian Council on Animal Care, 1993, p 92).
3. *Physical stress, such as physical or chair restraint, most definitely affects the behavior and psychology of laboratory animals. All possible measures to reduce their incidence should be taken. Animals should be trained to be as cooperative as possible to the procedures to facilitate the rapid completion of work and to alleviate stress in both the animals and people in charge* (Primate Research Institute, 2003, Chapter IV).
4. *Primates of many species can be quickly trained using positive reinforcement techniques to cooperate with a wide range of scientific, veterinary and husbandry*

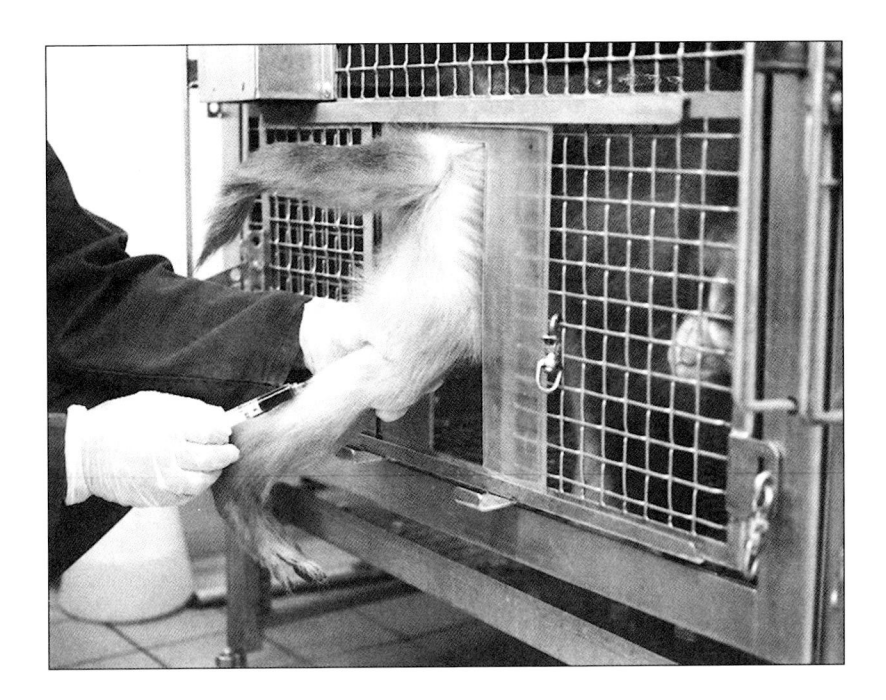

procedures. Such training is advocated whenever possible as a less stressful alternative to traditional methods using physical restraint. Techniques that reduce or eliminate adverse effects not only benefit animal welfare but can also enhance the quality of scientific research, since suffering in animals can result in physiological changes which are, at least, likely to increase variability in experimental data and, at worst, may even invalidate the research. Restraint procedures should be used only when less stressful alternatives are not feasible (International Primatological Society, 2007, p 22).

5. *The least distressing method of handling is to train the animal to cooperate in routine procedures. Advantage should be taken of the animal's ability to learn* (Home Office, 1989, p 18).
6. *Primates dislike being handled and are stressed by it; training animals to cooperate should be encouraged, as this will reduce the stress otherwise caused by handling. Training the animals is a most important aspect of husbandry, particularly in long-term studies. Training can often be employed to encourage the animals to accept minor interventions, such as blood sampling* (Council of Europe, 2006, p 48).

Despite these common-sense recommendations and the published fact that primates can readily be trained to cooperate during common handling procedures, there is resistance to implement positive reinforcement training as a standard refinement practice in biomedical research institutions. The reason for this inertia of tradition is probably related to misconceptions that have been published in text books and scientific articles:

1. ***All*** [emphasis added] *monkeys are dangerous* (Ackerley and Stones, 1969, p 207).
2. *Rhesus monkeys in the laboratory have well-earned reputations for their aggressive response and near-intractable disposition* (Bernstein et al., 1974, p 212).
3. *Old World primates **are** [emphasis added] aggressive and unpredictable* (IACUC Certification Coordinator, 2008, Web site).
4. *Nonhuman primates **are** [emphasis added] difficult and dangerous to handle* (Henrickson, 1976, p 62).
5. *One of the major drawbacks to the use of nonhuman primates is that they can be difficult and even dangerous to handle. Restraint is therefore **necessary** [emphasis added] and desirable to protect both the investigator and the animal* (Robbins et al., 1986, p 68).
6. *Primates can injure personnel severely if adequate restraint is not used. The risk of herpes virus B infection and other zoonoses transmitted by bite or scratch is minimized by appropriate restraint which may be physical or chemical or a combination of the two* (Whitney et al., 1973, p 50).
7. *Adult male rhesus monkeys **are** [emphasis added] aggressive animals and very difficult to handle. Hence experimental manipulations necessarily involve the use of restraint procedures, either chemical or physical* (Wickings and Nieschlag, 1980, p 287).
8. *Nonhuman primates, no matter how small, can be a danger to handlers. Restraint is **necessary** [emphasis added] to allow sample collection, drug administration or physical examination* (Panneton et al., 2001, p 92).

The reviewed literature suggests that these rather sweeping statements, albeit made by scientists, are based on beliefs rather than facts. That they are taken at face value by other scientists is regrettable as it promotes one of the most important extraneous variables, namely restraint stress. It is an irony that nonhuman primates are forcefully restrained in order to protect the handling personnel, yet *despite rigorous observance of all precautions, bites and scratches are frequent* (Valerio et al., 1969, p 45; cf. Zakaria et al., 1996; Sotir et al., 1997) because the animals are pushed into situations in which they have no other option but to defend themselves. When they have been trained to cooperate, they work with rather than against the handling personnel. Under these conditions handling procedures with primates are safe because the animals no longer have any reason to bite or scratch in self-defense.

The published reports on successful training protocols for injection, blood collection, semen collection, saliva collection, blood pressure measurement, oral drug administration, topical drug administration and weighing are encouraging. Their systematic application in the species for which they were originally developed, and their adaptation to other species will make the handling procedures with nonhuman primates more "humane" and the research data collected scientifically more valid.

6. CONCLUSIONS

The traditional housing and handling practices of caged primates expose the animals to unnecessary distress, which is not only an ethical concern—distress is a sign of impaired well-being—but also a scientific concern—distress is an uncontrolled variable that increases statistical variance.

It is documented in professional and scientific journals that housing and handling practices of caged nonhuman primates can be refined, without undue labor and expenses, in such a way that distress responses are minimized or avoided if basic ethological principles are applied to:
1. address the animal's need to be with and interact with at least one compatible conspecific;
2. structure their living quarters in species-appropriate ways;
3. address their biologically strong motivation to forage;
4. train them to cooperate during procedures.

With a little bit of good will and earnest concern for animal welfare and scientific methodology, the systematic implementation of Refinement for caged nonhuman primates is a practical option.

It must be remembered that the goal of Refinement is to decrease the incidence or severity of *inhumane* practices (Russell and Burch, 1992). The National Research Council (1985, p1) of the United States:
- *claims that the scientific community* [has] *long recognized both a scientific and an ethical responsibility for the* **humane** [emphasis added] *care of animals,* and
- *admonishes that all who care for or use animals in research, testing and education must assume responsibility for their general* **welfare** [emphasis added].

Is it *humane* and does it promote animal *welfare* when animals, who are known to have strong social needs, are kept alone in single-cages on a permanent basis?

Is it *humane* and does it promote animal *welfare* when animals, who show a biological vertical flight response, are permanently kept in cages without a high resting surface?

Is it *humane* and does it promote animal *welfare* when animals, who are highly motivated to engage in foraging behavior, receive their daily food ration in such a way that no effort is required to search, retrieve and process the food?

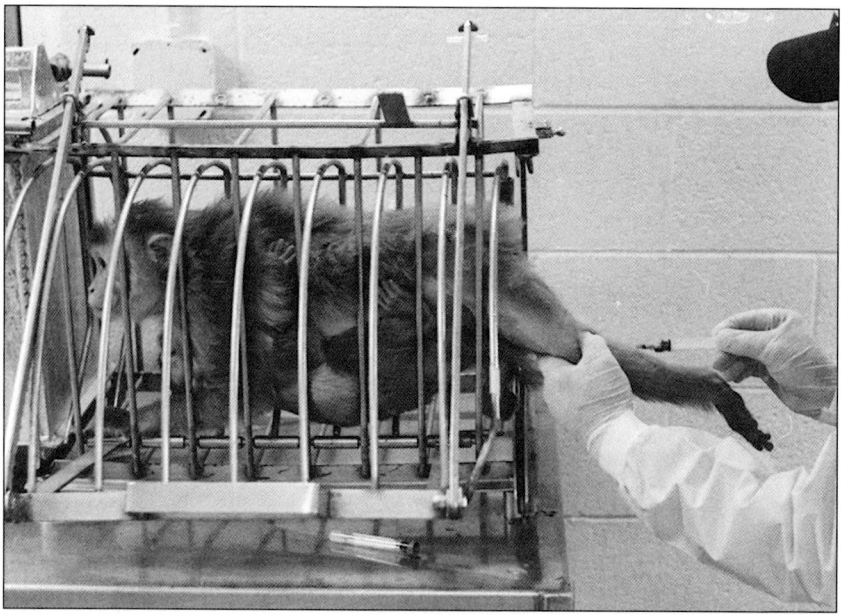

Is it *humane* and does it promote animal *welfare* when intelligent animals, who could readily learn how to cooperate, are forcefully restrained during common procedures?

7. REFERENCES

Abney D, Conlee K, Cunneen M, Down N, Lang T, Patterson-Kane E, Skoumbourdis E and Reinhardt V 2006 Human-animal relationship in the research lab: a discussion by the Refinement and Enrichment Forum. *Animal Technology and Welfare 5*: 95-98
http://www.awionline.org/Lab_animals/biblio/atw11.html

Ackerley ET and Stones PB 1969 Safety procedures for handling monkeys. *Laboratory Animal Handbooks 4*: 207-211

Alexander S and Fontenot MB 2003 Isosexual social group formation for environmental enrichment in adult male *Macaca mulatta. AALAS [American Association for Laboratory Animal Science] 54th National Meeting Official Program*: 141

Allen K, Blascovich J, Tomaka J and Kelsey R 1991 Presence of human friends and pet dogs as moderators of autonomic responses to stress in women. *Journal of Personality and Social Psychology 61*: 582-589

Allen K, Shykoff B and Izzo JL 2001 Pet ownership, but not ACE inhibitor therapy, blunts home blood pressure responses to mental stress. *Hypertension 38*: 815-820
http://hyper.ahajournals.org/cgi/content/full/38/4/815?ijkey=4113b0eaa8f97b29ddc8833ff3060acdc1f719e2

Allen K, Blascovich J and Mendes W 2002 Cardiovascular reactivity and the presence of pets, friends, and spouses: the truth about cats and dogs. *Psychosomatic Medicine 64*: 727-737
http://www.psychosomaticmedicine.org/cgi/content/full/64/5/727?ijkey=8efe81652f59a18ee3f91a289fe92117c7d4b076

Allentown Caging Equipment 2002 *Cat No. GP02*. Allentown Caging Equipment Co., Inc.: Allentown, NJ
http://www.acecaging.com/Primate1.htm

American Association for Laboratory Animal Science 2001 *Cost of Caring: Recognizing Human Emotions in the Care of Laboratory Animals*. American Association for Laboratory Animal Science: Memphis, TN
http://www.aalas.org/pdf/06-00006.pdf

American Psychiatric Association 1987 *Diagnostic and Statistical Manual of Mental Disorders* (Third Edition). American Psychiatic Association: Washington, DC

Anchel M 1976 Beyond "adequate veterinary care." *Journal of the American Veterinary Medical Association 168*: 513-517

Andersen HS, Sestoft D, Lillebaek T, Gabrielsen G, Hemmingsen R and Kramp P 2000 A longitudinal study of prisoners on remand: psychiatric prevalence, incidence and psychopathology in solitary vs. non-solitary confinement. *Acta Psychiatrica Scandinavica 102*: 19-25

Andersen HS, Sestoft D, Lillebaek T, Gabrielsen G and Hemmingsen R 2003 A longitudinal study of prisoners on remand: repeated measures of psychopathology in the initial phase of solitary versus nonsolitary confinement. *International Journal of Law and Psychiatry 26*: 165-177

Anderson JR and Chamove AS 1981 Self-aggressive behaviour in monkeys. *Current Psychological Reviews 1*: 139-158

Anderson JH and Houghton P 1983 The pole and collar system. A technique for handling and training nonhuman primates. *Lab Animal 12*(6): 47-49

Anderson JR and Chamove AS 1984 Allowing captive primates to forage. *Standards in Laboratory Animal Management. Proceedings of a Symposium* pp. 253-256. Universities Federation for Animal Welfare: Potters Bar, UK

Anderson WP, Reid CM and Jennings GL 1992 Pet ownership and risk factors for cardiovascular disease. *Medical Journal of Australia 157*: 298-301

Anonymous 2004 Self-injurious biting in laboratory animals: A discussion. *Laboratory Primate Newsletter 43*(2): 11-13
http://www.brown.edu/primate/lpn43-2.html#sib

Arluke AB and Sanders CR 1996 *Regarding Animals*. Temple University Press: Philadelphia, PA

Arrigo BA and Bullock JL 2007 The psychological effects of solitary confinement on prisoners in Supermax Units. *International Journal of Offender Therapy and Comparative Criminology, in press*

Arsenian JM 1943 Young children in an insecure situation. *Journal of Abnormal and Social Psychology 38*: 225-249

Astrup CW, Gantt WH and Stephens JH 1979 Differential effects of person in the dog and in the human. *Pavlovian Journal of Biological Science 14*: 104-107

Bach-Rita G 1974 Habitual violence and self-mutilation: A profile of 62 men. *American Journal of Psychiatry 131*: 1018-1020

REFERENCES

Baker KC 1999 Affiliative interactions between singly-housed rhesus macaques in adjacent unmodified cages. *American Journal of Primatology 49*: 30

Baker KC 2007 Grooming between pair-housed adult male rhesus macaques. *Personal Communication*

Baker KC, Weed JL, Crockett CM and Bloomsmith MA 2007 Survey of environmental enhancement programs for laboratory primates. *American Journal of Primatology 69*: 377-394

Balls M 1995 The use of non-human primates as laboratory animals in Europe: Moving toward the zero option. *ATLA [Alternatives to Laboratory Animals] 23*: 284-286

Barker SB, Knisely JS, McCain NL and Best AM 2005 Measuring stress and immune response in healthcare professionals following interaction with a therapy dog: a pilot study. *Psychology Reports 96*: 713-729

Barnett JL, Hemsworth PH, Hennessy DP, McCallum TH and Newman EA 1994 The effects of modifying the amount of human contact on behavioural, physiological and production responses of laying hens. *Applied Animal Behaviour Science 41*: 87-100

Barrow S, Luschei E, Nathan M and Saslow C 1966 A training technique for the daily chairing of monkeys. *Journal of the Experimental Analysis of Behavior 9*: 680

Basile BM, Hampton RR, Chaudhry AM and Murray EA 2007 Presence of a privacy divider increases proximity in pair-housed rhesus monkeys. *Animal Welfare 16*: 37-39

Baumans V, Coke C, Green J, Moreau E, Morton D, Patterson-Kane E, Reinhardt A, Reinhardt V and Van Loo P (eds) 2007 *Making Lives Easier for Animals in Research Labs: Discussions by the Laboratory Animal Refinement & Enrichment Forum.* Animal Welfare Institute: Washington, DC
http://www.awionline.org/pubs/LAREF/LAREF-bk.html

Baun MM, Bergstrom N, Langston NF and Thoma L 1984 Physiological effects of human/companion animal bonding. *Nursing Research 33*: 126-129

Bayne K 2002 Development of the human-research animal bond and its impact on animal well-being. *ILAR [Institute for Laboratory Animal Research] Journal 43*: 4-9
http://dels.nas.edu/ilar_n/ilarjournal/43_1/Development.shtml

Bayne K, Dexter SL, Mainzer H, McCully C, Campbell G and Yamada F 1992a The use of artificial turf as a foraging substrate for individually housed rhesus monkeys *(Macaca mulatta)*. *Animal Welfare 1*: 39-53
http://www.awionline.org/Lab_animals/biblio/aw1-39.htm

Bayne K, Hurst JK and Dexter SL 1992b Evaluation of the preference to and behavioral effects of an enriched environment on male rhesus monkeys. *Laboratory Animal Science 42*: 38-45

Bayrakci R 2003 Starting an injection training program with lion-tailed macaques (*Macaca silenus*). *Animal Keeper's Forum 30*: 503-512

Bentson KL, Capitanio JP and Mendoza SP 2003 Cortisol responses to immobilization with Telazol or ketamine in baboons (*Papio cynocephalus/anubis*) and rhesus macaques (*Macaca mulatta*). *Journal of Medical Primatology 32*: 148-160

Berendt R and Williams TD 1971 The effect of restraint and position upon selected respiratory parameters of two species of *Macaca*. *Laboratory Animal Science 21*: 502-509

Berkman LF 1985 The relationship of social netwworks and social support to morbidity and mortality. In: Cohen S and Syme SL (eds) *Social Support and Health* pp. 241-262. Academic Press: Orlando, FL

Bernstein IS and Mason WA 1963 Group formation by rhesus monkeys. *Animal Behaviour 11*: 28-31

Bernstein IS, Gordon TP and Rose RM 1974 Factors influencing the expression of aggression during introductions to rhesus monkey groups. In: Holloway RL (ed) *Primate Aggression, Territoriality, and Xenophobia* pp. 211-240. Academic Press: New York, NY

Bertrand F, Seguin Y, Chauvier F and Blanquié JP 1999 Influence of two different kinds of foraging devices on feeding behaviour of rhesus macaques *(Macaca mulatta)*. *Folia Primatologica 70*: 207

Biondi M and Picardi A 1996 Clinical and biological aspects of bereavement and loss-induced depression: a reappraisal. *Psychotherapy and Psychosomatics 65*: 229-245

Bjone SJ, Price IR and McGreevy PD 2006 Food distribution effects on the behaviour of captive common marmosets, *Callithrix jacchus*. *Animal Welfare 15*: 131-140

Bland M, Bridge C, Cooper M, Dixon D, Hay L and Zerbato A 2002 Procedural restraint in children's nursing: using clinical benchmarks. *Professional Nurse 17*: 712-715

Boinski S, Gross TS and Davis JK 1999 Terrestrial predator alarm vocalizations are a valid monitor of stress in captive brown capuchins *(Cebus apella)*. *Zoo Biology 18*: 295-312

Bonnie KE and de Waal FB 2006 Affiliation promotes the transmission of a social custom: handclasp grooming among captive chimpanzees. *Primates 47*: 27-34

Bourgeois SR and Brent L 2005 Modifying the behaviour of singly caged baboons: evaluating the effectiveness of four enrichment techniques. *Animal Welfare 14*: 71-81

Bowers CL, Crockett CM and Bowden DM 1998 Differences in stress reactivity of laboratory macaques measured by heart period and respiratory sinus arrhythmia. *American Journal of Primatology 45*: 245-261

Boyce W, Champoux M, Suomi S and Gunnar M 1995 Salivary cotisol in nursery-reared rhesus monkeys: reactivity to peer interactions and altered circadian activity. *Developmental Psychology 28*: 257-267

Brenner M 2007 Child restraint in the acute setting of pediatric nursing: an extraordinarily stressful event. *Issues in Comprehensive Pediatric Nursing 30*(1-2): 29-37

Broadhead WE, Kaplan BH, James SA, Wagner EH, Schoenbach VJ, Grimson R, Heyden S, Tibblin G and Gehlbach SH 1983 The epidemiologic evidence for a relationship between social support and health. *American Journal of Epidemiology 117*: 521-537

Brockway BP, Hassler CR and Hicks N 1993 Minimizing stress during physiological monitoring. In: Niemi SM and Willson JE (eds) *Refinement and Reduction in Animal Testing* pp. 56-69. Scientists Center for Animal Welfare: Bethesda, MD

Brown, CS 1998 *A Training Program for Semen Collection in Gorillas* (Videotape with commentary). Omaha's Henry Doorly Zoo: Omaha, NB

Brown CS and Loskutoff NM 1998 A training program for noninvasive semen collection in captive western lowland gorillas *(Gorilla gorilla gorilla)*. *Zoo Biology 17*: 143-151

Bryant CE, Rupniak NMJ and Iversen SD 1988 Effects of different environmental enrichment devices on cage stereotypies and autoaggression in captive cynomolgus monkeys. *Journal of Medical Primatology 17*: 257-269
http://www.awionline.org/Lab_animals/biblio/jmp17-2.htm

Buchanan-Smith HM, Rennie AE, Vitale A, Pollo S, Prescott MJ and Morton DB 2005 Harmonising the definition of refinement. *Animal Welfare 14*: 379-384

Byrd LD 1977 Introduction: Chimpanzees as biomedical models. In: Bourne GH (ed) *Progress in Ape Research* pp. 161-165. Academic Press: New York, NY

Byrum R and St. Claire M 1998 Pairing female *Macaca nemestrina*. *Laboratory Primate Newsletter 37*(4): 1
http://www.brown.edu/Research/Primate/lpn37-4.html#byrum

Capitanio JP, Mendoza SP and McChesney M 1996 Influence of blood sampling procedures on basal hypothalamic-pituitary-adrenal hormone levels and leukocyte values in rhesus macaques *(Macaca mulatta)*. *Journal of Medical Primatology 25*: 26-33
http://www.awionline.org/Lab_animals/biblio/jmp25-2.htm

Canadian Council on Animal Care 1984 Chapter XX: Non-human primates. In: *Guide to the Care and Use of Experimental Animals,* Volume 2 pp. 163-173. Canadian Council on Animal Care: Ottawa, Canada
http://www.ccac.ca/en/CCAC_Programs/Guidelines_Policies/GUIDES/ENGLISH/V2_84/CHXX.HTM

Canadian Council on Animal Care 1993 *Guide to the Care and Use of Experimental Animals,* Volume 1 (Second Edition). Canadian Council on Animal Care: Ottawa, Canada
http://www.ccac.ca/en/CCAC_Programs/Guidelines_Policies/GUIDES/ENGLISH/toc_v1.htm

Casanova CCN and Garcia MSJ 1996 Hierarchy of social "friendship" and hierarchy of social dominance in captive chimpanzees (*Pan troglodytes*). *XVIth Congress of the International Primatological Society/XIXth Conference of the American Society of Primatologists*: 329

Celli ML, Tomonagaa M, Udonob T, Teramotob M and Naganob K 2003 Tool use task as environmental enrichment for captive chimpanzees. *Applied Animal Behaviour Science 81*: 171-182

Chambers DR, Gibson TE, Bindman L, Guillou PJ, Herbert WJ, Mayes PA, Poole TB, Wade AJ, Wood RKS [The Biological Council Animal Research and Welfare Panel] 1992 *Guidelines on the Handling and Training of Laboratory Animals*. Universities Federation for Animal Welfare: Potters Bar, UK

Chamove AS 1989 Environmental enrichment: A review. *Animal Technology 40*: 155-178
http://www.awionline.org/Lab_animals/biblio/at89-cham.html

Chamove AS 2001 Floor-covering research benefits primates. *Australian Primatology 14*(3): 16-19
http://www.lisp.com.au/~primate/arnold.htm

Chamove AS and Scott L 2005 Forage box as enrichment in single- and group-housed callitrichid monkeys. *Laboratory Primate Newsletter 44*(2): 13-17
http://www.brown.edu/Research/Primate/lpn44-2.html#box

Chance MRA 1956 Social structure of a colony of *Macaca mulatta*. *Animal Behaviour 4*: 1-13

Chance MRA 1961 The nature and special features of the instinctive social bond of primates. *Viking Fund Publications in Anthropology 31*: 17-34

REFERENCES

Chance MRA 1975 Social cohesion and the structure of attention. In: Fox R (ed) *Biosocial Anthropology* pp. 93-113. John Wiley and Sons: New York, NY

Chance, MRA and Jolly, CJ 1970 *Social Groups of Monkeys, Apes and Men*. E. P. Dutton and Company, Inc.: New York, NY

Chopra PK, Seth PK and Seth S 1992 Behavioural profile of free-ranging rhesus monkeys. *Primate Report 32*: 75-105

Christenfeld N, Gerin W, Linden W, Sanders M, Mathur J, Deich JD and Pickering TG 1997 Social support effects on cardiovascular reactivity: is a stranger as effective as a friend? *Psychosomatic Medicine 59*: 388-398
http://www.psychosomaticmedicine.org/cgi/reprint/59/4/388?ijkey=7bb0a15feedd76fedfdaad078442f9d73fe8b24f

Christenfeld N and Gerin W 2000 Social support and cardiovascular reactivity. *Biomedicine and Pharmacotherapy 54*: 251-257

Christenson GA and Mansueto CS 1999 Trichotillomania: descriptive characteristics and phenomenology. In: Stein DJ, Christenson GA and Hollander E (eds) *Trichotillomania* pp. 1-41. American Psychiatric Press: Washington, DC

Clarence WM, Scott JP, Dorris MC and Paré M 2006 Use of Enclosures with Functional Vertical Space by Captive Rhesus Monkeys (*Macaca mulatta*) Involved in Biomedical Research. *Journal of the American Association for Laboratory Animal Science [Contemporary Topics in Laboratory Animal Science] 45*(5): 31-34

Clarke AS, Mason WA and Moberg GP 1988 Interspecific contrasts in responses of macaques to transport cage training. *Laboratory Animal Science 38*: 305-309

Clarke AS, Czekala NM and Lindburg DG 1995 Behavioral and adrenocortical responses of male cynomolgus and lion-tailed macaques to social stimulation and group formation. *Primates 36*: 41-46

Clarke MR and Blanchard JL 1994 All-male social group formation: Does cutting canine teeth promote social integration? *Laboratory Primate Newsletter 33*(2): 5-8
http://www.brown.edu/Research/Primate/lpn33-2.html#clarke

Coe CL 1991 Is social housing of primates always the optimal choice? In: Novak MA and Petto AJ (eds) *Through the Looking Glass. Issues of Psychological Well-being in Captive Nonhuman Primates* pp. 78-92. American Psychological Association: Washington, DC

Coe CL, Franklin D, Smith ER and Levine S 1982 Hormonal responses accompanying fear and agitation in the squirrel monkey. *Physiology and Behavior 29*: 1051-1057

Coelho AM and Carey KD 1990 A social tethering system for non-human primates used in laboratory research. *Laboratory Animal Science 40*: 388-394

Coelho AM, Carey KD and Shade RE 1991 Assessing the effects of social environment on blood pressure and heart rates of baboons. *American Journal of Primatology 23*: 257-267

Cohen, S and Syme, SL 1985 *Social Support and Health*. Academic Press: San Diego, CA

Coke CS, Priddy E, Wright J, Salleng K and Wallace J 2007 Introducing macaques to novel housing systems mid-experiment: Proactive behavioral management. *AALAS [American Association for Laboratory Animal Science] 58th National Meeting Official Program*: 53

Cole KM, Gawlinski A, Steers. M. and Kotlerman J 2007 Animal-assisted therapy in patients hospitalized with heart failure. *American Journal of Critical Care 16*: 575-585 http://ajcc.aacnjournals.org/cgi/content/full/16/6/575

Coppola CL, Grandin T and Enns RM 2006 Human interaction and cortisol: can human contact reduce stress for shelter dogs? *Physiology and Behavior 87*: 537-541

Council of Europe 2006 *Progressive regulations pertaining to the species-appropriate housing and handling of animals kept in research labs. Appendix A of the European Convention for the Protection of Vertebrate Animals Used for Experimental and Other Scientific Purposes (ETS No. 123)*. Council of Europe: Strasbourg, France http://conventions.coe.int/Treaty/EN/Treaties/PDF/123-Arev.pdf

Craig FW, Lynch JJ and Quartner JL 2000 The perception of available social support is related to reduced cardiovascular reactivity in Phase II cardiac rehabilitation patients. *Integrative Physiological and Behavioral Science 35*: 272-283

Crockett CM 1993 Rigid rules for promoting psychological well-being are premature. *American Journal of Primatology 30*: 177-179

Crockett CM and Bowden DM 1994 Challenging conventional wisdom for housing monkeys. *Lab Animal 24*(2): 29-33

Crockett CM and Heffernan KS 1998 Grooming-contact cages promote affiliative social interaction in individually housed adult baboons. *American Journal of Primatology 45*: 176

Crockett C. M. and Gough GM 2002 Onset of aggressive toy biting by a laboratory baboon coincides with cessation of self-injurious behavior. *American Journal of Primatology 57*: 39 http://www.asp.org/asp2002/abstractDisplay.cfm?abstractID=306&confEventID=285

Crockett CM, Bowers CL, Bowden DM and Sackett GP 1994 Sex differences in compatibility of pair-housed adult longtailed macaques. *American Journal of Primatology 32*: 73-94

Crockett CM, Bowers CL, Shimoji M, Leu M, Bowden DM and Sackett GP 1995 Behavioral responses of longtailed macaques to different cage sizes and common laboratory experiences. *Journal of Comparative Psychology 109*: 368-383

Crockett CM, Bellanca RU, Bowers CL and Bowden DM 1997 Grooming-contact bars provide social contact for individually caged laboratory primates. *Contemporary Topics in Laboratory Animal Science 36*(6): 53-60

Crockett CM, Koberstein D and Heffernan KS 2001 Compatibility of laboratory monkeys housed in grooming-contact cages varies by species and sex. *American Journal of Primatology 54*(Supplement): 51-52
http://www.asp.org/asp2001/abstractDisplay.cfm?abstractID=112&confEventID=26

Crockett CM, Lee GH and Thom JP 2006 Sex and age predictors of compatibility in grooming-contact caging vary by species of laboratory monkey. *International Journal of Primatology 27*(Supplement): 417

Cross N, Pines MK and Rogers LJ 2004 Saliva sampling to assess cortisol levels in unrestrained common marmosets and the effect of behavioral stress. *American Journal of Primatology 62*: 107-114

Davenport MD, Lutz CK, Tiefenbacher S, Novak MA and Meyer JS 2007 A rhesus monkey model of self-injury: Effects of relocation stress on behavior and neuroendocrine function. *Biological Psychiatry, in press*

de Rosa C, Vitale A and Puopolo M 2003 The puzzle-feeder as feeding enrichment for common marmosets (*Callithrix jacchus*): a pilot study. *Laboratory Animals 37*: 100-107

de Waal FBM and Luttrell LM 1986 The similarity principle underlying social bonding among female rhesus monkeys. *Folia Primatologica 46*: 215-234

de Waal FBM and Ren RM 1988 Comparison of the reconciliation behavior of stumptail and rhesus macaques. *Ethology 78*: 129-142

Dean SW 1999 Environmental enrichment of laboratory animals used in regulatory toxicology studies. *Laboratory Animals 33*: 309-327

Dettmer E and Fragaszy DM 2000 Determining the value of social companionship to captive tufted capuchin monkeys *(Cebus apella)*. *Journal of Applied Animal Welfare Science 3*: 393-304

Dexter SL and Bayne K 1994 Results of providing swings to individually housed rhesus monkeys (*Macaca mulatta*). *Laboratory Primate Newsletter 33*(2): 9-12
http://www.brown.edu/Research/Primate/lpn33-2.html#bayne

Down N, Skoumbourdis E, Walsh M, Francis R, Buckmaster C and Reinhardt V 2005 Pole-and-collar training: A discussion by the Laboratory Animal Refinement & Enrichment Forum. *Animal Technology and Welfare 4*: 157-161
http://www.awionline.org/Lab_animals/biblio/atw7.html

Doyle LA, Baker KC and Cox LD 2008 Physiological and behavioral effects of social introduction on adult male rhesus macaques. *American Journal of Primatology 70*: 1-9

Drescher VM, Gantt WH and Whitehead WE 1980 Heart rate response to touch. *Psychosomatic Medicine 42*: 559-559

Duffy KG 2006 The form and function of social bonds among wild male chimpanzees (Pan troglodytes schweinfurthii) in the Kibale National Park, Uganda. *International Journal of Primatology 27*(Supplement): 533

Eaton GG, Kelley ST, Axthelm MK, Iliff-Sizemore SA and Shiigi SM 1994 Psychological well-being in paired adult female rhesus (*Macaca mulatta*). *American Journal of Primatology 33*: 89-99

Elvidge H, Challis JRG, Robinson JS, Roper C and Thorburn GD 1976 Influence of handling and sedation on plasma cortisol in rhesus monkeys *(Macaca mulatta)*. *Journal of Endocrinology 70*: 325-326

Epley SW 1974 Reduction of the behavioral effects of aversive stimulation by the presence of companions. *Psychological Bulletin 81*: 271-283

Erkert HG 1999 Owl monkeys. In: UFAW [Universities Federation for Animal Welfare] (edited by Poole, T. and English, P.) *The UFAW Handbook on the Care and Management of Laboratory Animals Seventh Edition* pp. 585-590. Blackwell Science: Oxford, UK

Erwin J 1979 Aggression in captive macaques: Interaction of social and spacial factors. In: Erwin J, Maple T and Mitchell G (eds) *Captivity and Behavior* pp. 139-171. Van Nostrand: New York, NY

Erwin J 1991 Applied primate ecology: Evaluation of environmental changes to promote psychological well-being. In: Novak MA and Petto AJ (eds) *Through the Looking Glass. Issues of Psychological Well-Being in Captive Nonhuman Primates* pp. 180-188. American Psychological Association: Washington, DC

Erwin J and Deni R 1979 Strangers in a strange land: Abnormal behavior or abnormal environments? In: Erwin J, Maple T and Mitchell G (eds) *Captivity and Behavior* pp. 1-28. Van Nostrand Reinhold: New York, NY

Erwin J, Mitchell G and Maple T 1973 Abnormal behavior in non-isolate-reared rhesus monkeys. *Psychological Reports 33*: 515-523

REFERENCES

European Commission 2002 *The Welfare of Non-human Primates - Report of the Scientific Committee on Animal Health and Animal Welfare*. European Commission: Strasbourg, France http://europa.eu.int/comm/food/fs/sc/scah/out83_en.pdf

Evans HL, Taylor JD, Ernst J and Graefe JF 1989 Methods to evaluate the well-being of laboratory primates. Comparison of macaques and tamarins. *Laboratory Animal Science 39*: 318-323

Fekete JM, Norcross JL and Newman JD 2000 Artificial turf foraging boards as environmental enrichment for pair-housed female squirrel monkeys. *Contemporary Topics in Laboratory Animal Science 39*(2): 22-26

Folkes K 2005 Is restraint a form of abuse? *Paediatric Nursing 17*(6): 41-44

Fontana AM, Diegnan T, Villeneuve A and Lepore S 1999 Nonevaluative social support reduces cardiovascular reactivity in young women during acutely stressful performance situations. *Journal of Behavioral Medicine 22*: 75-91

Fowler, ME 1995 *Restraint and Handling of Wild and Domestic Animals* (Second Edition). Iowa State University Press: Ames, IA

Friedmann E and Thomas SA 1995 Pet ownership, social support, and one-year survival after acute myocardial infarction in the Cardiac Arrhythmia Suppression Trial (CAST). *American Journal of Cardiology 76*: 1213-1217

Friedmann E, Katcher AH, Lynch JJ and Thomas SA 1980 Animal companions and one-year survival of patients after discharge from a coronary care unit. *Public Health Reports 95*: 307-312

Friedmann E, Katcher AH, Thomas SA, Lynch JJ and Messent PR 1983 Social interaction and blood pressure: influence of animal companions. *Journal of Nervous and Mental Disease 171*: 461-454
http://www.pubmedcentral.nih.gov/picrender.fcgi?artid=1422527&blobtype=pdf

Friscino BH, Gai CL, Kulick AA, Donnelly MJ, Rockar RA, Aderson LC and Iliff SA 2003 Positive reinforcement training as a refinement of a macaque biliary diversion model. *AALAS [American Association for Laboratory Animal Science] 54th National Meeting Official Program*: 101

Fritz J 1989 Resocialization of captive chimpanzees: An amelioration procedure. *American Journal of Primatology 19*(Supplement): 79-86

Fritz J 1994 Introducing unfamiliar chimpanzees to a group or partner. *Laboratory Primate Newsletter 33*(1): 5-7
http://www.brown.edu/Research/Primate/lpn33-1.html#jo

Fritz P and Fritz J 1979 Resocialization of chimpanzees: Ten years of experience at the Primate Foundation of Arizona. *Journal of Medical Primatology 8*: 202-221

Fruth B and Hohmann G 1998 Spatial association and social affiliation among female bonobos (*Pan paniscus*). *Folia Primatologica 69*: 198

Fujisawa K, Kutsukake N and Hasegawa T 2004 Similarity in conflict resolution pattern between primates and preschool children: reconciliation form and friendship. *Folia Primatologica 55* (Supplement): 377

Gallo LC, Smith TW and Kircher JC 2000 Cardiovascular and electrodermal responses to support and provocation: interpersonal methods in the study of psychophysiological reactivity. *Psychophysiology 37*: 289-301

Gamman T 1995 Hazardous health effect of isolation. A clinical study of 2 groups of persons in custody [article in Norwegian with English abstract]. *Tidsskrift for den Norske Lægeforening 115*: 2243

Gantt WH, Newton JEO, Royer FL and Stephens JH 1966 Effect of person. *Conditional Reflex 1*: 18-35

Gerber P, Schnell CR and Anzenberger G 2002 Behavioral and cardiophysiological responses of common marmosets (*Callithrix jacchus*) to social and environmental changes. *Primates 43*: 201-216

Gerin W, Pieper C, Levy R and Pickering TG 1992 Social support in social interaction: a moderator of cardiovascular reactivity. *Psychosomatic Medicine 54*: 324-326
http://www.psychosomaticmedicine.org/cgi/reprint/57/1/16?ijkey=1d0734a564bd72f2109d781226af979872898973

Gerin W, Milner D, Chawla S and Pickering TG 1995 Social support as a moderator of cardiovascular reactivity in women: A test of the direct effects and buffering hypothesis. *Psychosomatic Medicine 57*: 16-22
http://www.psychosomaticmedicine.org/cgi/reprint/57/1/16

Gilbert SG and Wrenshall E 1989 Environmental enrichment for monkeys used in behavioral toxicology studies. In: Segal EF (ed) *Housing, Care and Psychological Well-Being of Captive and Laboratory Primates* pp. 244-254. Noyes Publications: Park Ridge, NJ

Gisler DB, Benson RE and Young RJ 1960 Colony husbandry of research monkeys. *Annals of the New York Academy of Sciences 85*: 758-568

Glick-Bauer M 1997 Behavioral enrichment for captive cotton-top tamarins (*Saguinus oedipus*) through novel presentation of diet. *Laboratory Primate Newsletter 36*(1): 1-3
http://www.brown.edu/Research/Primate/lpn36-1.html#glick

Gluck J and Sackett G 1974 Frustration and self-aggression in social isolate rhesus monkeys. *Journal of Abnormal Psychology 83*: 331-334

Golub MS and Anderson JH 1986 Adaptation of pregnant rhesus monkeys to short-term chair restraint. *Laboratory Animal Science 36*: 507-511

Gonzalez CA, Coe CL and Levine S 1982 Cortisol responses under different housing conditions in female squirrel monkeys. *Psychoneuroendocrinology 7*: 209-216

Goodall J 1964 Tool-using and aimed throwing in a community of free-living chimpanzees. *Nature 201*: 1264-1266

Goosen C, van der Gulden W and Rozemond H 1984 Recommendations for the housing of macaque monkeys. *Laboratory Animals 18*: 245-249

Grant JL and Doudet DJ 2003 Obtaining blood samples from awake rhesus monkeys (*Macaca mulatta*). *Laboratory Primate Newsletter 42*(2): 1-3
http://www.brown.edu/Research/Primate/lpn42-1.html#blood

Grassian S 1983 Psychopathological effects of solitary confinement. *American Journal of Psychiatry 140*: 1450-1454

Grassian S and Friedman N 1986 Effects of sensory deprivation in psychiatric seclusion and solitary confinement. *International Journal of Law and Psychiatry 8*: 49-65

Gunnar MR, Gonzalez CA and Levine S 1980 The role of peers in modifying behavioral distress and pituitary-adrenal response to a novel environment in year-old rhesus monkeys. *Physiology and Behavior 25*: 795-798

Gust DA, Gordon TP, Brodie AR and McClure HM 1994 Effect of a preferred companion in modulating stress in adult female rhesus monkeys. *Physiology and Behavior 55*: 681-684

Gust DA, Gordon TP, Wilson ME, Brodie AR, Ahmed-Ansari A and McClure HM 1996 Group formation of female pigtail macaques (*Macaca nemestrina*). *American Journal of Primatology 39*: 263-273

Gwinn LA 1996 A method for using a pole housing apparatus to establish compatible pairs among squirrel monkeys. *Contemporary Topics in Laboratory Animal Science 35*(4): 61

Hall KRL and De Vore I 1965 Baboon social behavior. In: De Vore I (ed) *Primate Behavior - Field Studies of Monkeys and Apes* pp. 53-110. Holt, Rinehart and Winston: New York, NY

Hallopeau H 1894 Sur un nouveau cas de trichotillomanie. *Annales de Dermatologie et de Syphiligraphie 5*: 541-543

Hamiel D, Yoffe A and Roe D 1999 Trichotillomania and the mourning process: a case report and review of the psychodynamics. *Israel Journal of Psychiatry & Related Sciences 36*: 192-199

Hanya G 2004 Seasonal variations in the activity budget of Japanese macaques in the coniferous forest of Yakushima: Effects of food and temperature. *American Journal of Primatology 63*: 164-177

Hassler CR, Moutvic RR, Hobson DW, Lordo RA, Vinci LT, Dill GS, Joiner RL and Hamlin RL 1989 Long-term arrhythmia analysis of primates pretreated with pyridostigmine, challenged with soman, and treated with atropine and 2-PAM. *Proceedings of the 1989 Medical Defense Bioscience Review* pp. 479-482. Johns Hopkins University: Columbia, MD

Hays JC, Steffens DC, Flint EP, Bosworth HB and George LK 2001 Does social support buffer functional decline in elderly patients with unipolar depression? *American Journal of Psychiatry 158*: 1850-1855

Heath M 1989 The training of cynomolgus monkeys and how the human/animal relationship improves with environmental and mental enrichment. *Animal Technology 40*: 11-22 http://www.awionline.org/Lab_animals/biblio/at40heath.html

Hein PR, Schatorje JS, Frencken HJ, Segers MF and Thomas CM 1989 Serum hormone levels in pregnant cynomolgus monkeys. *Journal of Medical Primatology 18*: 133-142

Hemelrijk CK, Meier. C. and Martin RD 1999 'Friendship' for fitness in chimpanzees? *Animal Behaviour 58*: 1223-1229

Hemsworth PH, Barnett JL and Hansen C 1981 The influence of handling by humans on the behaviour, growth and corticosteroids in the juvenile female pig. *Hormones and Behavior 15*: 396-403

Hennessy MB 1984 Presence of companion moderates arousal of monkeys with restricted social experience. *Physiology and Behavior 33*: 393-398

Hennessy MB 1997 Hypothalamic-pituitary-adrenal response to brief social separation. *Neuroscience and Biobehavioral Reviews 21* : 11-29

Henrickson RV 1976 The nonhuman primate. *Lab Animal 5*(4): 60-62

Hermano-Silva C and Lee PC 2004 Conflict and friendship in immature Guinea baboons. *Folia Primatologica 75*(Supplement): 273

Hernándes-López L, Mayagoitia L, Esquivel-Lacroix C, Rojas-Maya S and Mondragón-Ceballos R 1998 The menstrual cycle of the spider monkey *(Ateles geoffroyi)*. *American Journal of Primatology 44*: 183-195

Herndon JG, Turner JJ, Perachio AA, Blank MS and Collins DC 1984 Endocrine changes induced by venipuncture in rhesus monkeys. *Physiology and Behavior 32*: 673-676

Home Office 1989 *Animals (Scientific Procedures) Act 1986. Code of Practice for the Housing and Care of Animals Used in Scientific Procedures*. Her Majesty's Stationery Office: London, UK
http://scienceandresearch.homeoffice.gov.uk/animal-research/legislation/

Honess P, Gimpel J, Wolfensohn S and Mason G 2005 Alopecia scoring: the quantitative assessment of hair loss in captive macaques. *ATLA [Alternatives to Laboratory Animals] 33*: 193-206

House JS, Robbins C and Metznere HL 1982 The association of social relationships and activities with mortality: Prospective evidence from the Tecumsseh community health study. *American Journal of Epidemiology 116*: 123-140

House JS, Landis KR and Umberson D 1988 Social relationships and health. *Science 241*: 540-544

Hrapkiewicz, K, Medina, L and Holmes, DD 1998 *Clinical Medicine of Small Mammals and Primates* (Second Edition). Manson Publishing: London, UK

IACUC Certification Coordinator 2008 Restraint. *IACUC Learning Module - Primates* (Web site): Updated 01/02/2008
http://www.iacuc.arizona.edu/training/primate/rest.html

Iliff SA, Friscino BH and Anderson LC 2004 Refinement of study design using positive reinforcement training in macaques. *Folia Primatologica 75*(Supplement): 282-283

Institute for Laboratory Animal Research 1992 *Recognition and alleviation of pain and distress in laboratory animals*. National Academy Press: Washington, DC
http://books.nap.edu/books/0309042755/html/index.html

International Primatological Society 1989 International guidelines for the acquisition, care and breeding of nonhuman primates. *Primate Report 25*: 3-27

International Primatological Society 1993 IPS International guidelines for the acquisition, care and breeding of nonhuman primates, Codes of Practice 1-3. *Primate Report 35*: 3-29
http://pin.primate.wisc.edu/ips/codes.txt

International Primatological Society 2007 *IPS International Guidelines for the Acquisition, Care and Breeding of Nonhuman Primates*. International Primatological Society: Bronx, NY
http://www.internationalprimatologicalsociety.org/docs/IPS_International_Guidelines_for_the_Acquisition_Care_and_Breeding_of_Nonhuman_Primates_Second_Edition_2007.pdf

Ives M and Dack GM 1956 "Alarm reaction" and normal blood picture in *Macaca mulatta*. *Journal of Laboratory Clinical Medicine 47*: 723-729

Jackson MJ 2001 Environmental enrichment and husbandry of the MPTP-treated common marmoset. *Animal Technology 52*: 21-28

Jaeckel J 1988 The benefits of training rhesus monkeys living under laboratory conditions. In: Poole TB (ed) *Laboratory Animal Welfare Research: Primates* pp. 23-25. Universities Federation for Animal Welfare: Potters Bar, UK

Jensen GD, Blanton FL and Gribble DH 1980 Older monkeys' *(Macaca radiata)* response to new group formation: Behavior, reproduction and mortality. *Experimental Gerontology 15*: 399-406

Jorgensen MJ, Kinsey JH and Novak MA 1998 Risk factors for self-injurious behavior in captive rhesus monkeys *(Macaca mulatta)*. *American Journal of Primatology 45*: 187

Kamarck TW, Manuck SB and Jennings JR 1990 Social support reduces cardiovascular reactivity to psychological challenge: a laboratory model. *Psychosomatic Medicine 52*: 42-48
http://www.psychosomaticmedicine.org/cgi/reprint/52/1/42?maxtoshow=&HITS=10&hits=10&RESULTFORMAT=&searchid=1&FIRSTINDEX=10&minscore=5000&resourcetype=HWCIT

Kaplan BH, Cassel JC and Gore S 1977 Social support and health. *Medical Care 15*(Supplement): 47-58

Kapsalis E and Johnson RL 2006 Getting to know you: female-female consortships in free-ranging rhesus monkeys. In: Sommer V and Vasey PL (eds) *Homosexual Behaviour in Animals: an Evolutionary Perspective* pp. 220-237. Cambridge University Press: New York, NY

Kenney C, Ruppenthal G, Novak M and Suomi S 2006 Animal-operated foldable perch usage in rhesus macaques *(Macaca mulatta)*. *American Journal of Primatology 68*(Supplement): 60

Kissel S 1965 Stress-reducing properties of social stimuli. *Journal of Personality and Social Psychology 2*: 378-384

Kirschbaum C, Klauer T, Filipp SH and Hellhammer DH 1995 Sex-specific effects of social support on cortisol and subjective responses to acute psychological stress. *Psychosomatic Medicine 57*: 23-31

Kitchen AM and Martin AA 1996 The effects of cage size and complexity on the behaviour of captive common marmosets, *Callithrix jacchus jacchus*. *Laboratory Animals 30*: 317-326

REFERENCES

Klein HJ and Murray KA 1995 Part C. Restraint. In: Bennett BT, Abee CR and Henrickson R (eds) *Nonhuman Primates in Biomedical Research - Biology and Management* pp. 286-297. Academic Press: New York, NY

Ko SM 1999 Under-diagnosed psychiatric syndrome. I: Trichotillomania. *Annals of the Academy of Medicine Singapore 28*: 279-281

Koban TL, Miyamoto M, Donmoyer G and Hammar A 2005 Effects of positive reinforcement training on cortisol, hematology and cardiovascular parameters in cynomolgus macaques (*Macaca fascicularis*). *American Journal of Primatology 66*(Supplement): 148

Kopecky J and Reinhardt V 1991 Comparing the effectiveness of PVC swings versus PVC perches as environmental enrichment objects for caged female rhesus macaques. *Laboratory Primate Newsletter 30*(2): 5-6
http://www.brown.edu/Research/Primate/lpn30-2.html#vik

Kummer, H 1968 *Social Organization of Hamdryas Baboons*. Karger, Switzerland: Basel

Lam K, Rupniak NMJ and Iversen SD 1991 Use of a grooming and foraging substrate to reduce cage stereotypies in macaques. *Journal of Medical Primatology 20*: 104-109
http://www.awionline.org/Lab_animals/biblio/jmp20-1.htm

Laule G and Whittaker M 2001 The use of positive reinforcement techniques with chimpanzees for enhanced care and welfare. In: Brent L (ed.) *Special Topics in Primatology - The Care and Management of Captive Chimpanzees* pp. 242-266. The American Society of Primatologists: San Antonio, TX

Laule GE, Thurston RH, Alford PL and Bloomsmith MA 1996 Training to reliably obtain blood and urine samples from a diabetic chimpanzee (*Pan troglodytes*). *Zoo Biology 15*: 587-591

Lee GH, Thom JP and Crockett CM 2005 Factors predicting compatible grooming-contact pairings in four species of laboratory monkeys. *American Journal of Primatology 66* (Supplement): 83-84

Leon S and Taylor L 1993 Activity patterns of captive and free-ranging long-tailed macaques (*Macaca fascicularis*). *American Zoo and Aquarium Association (AZA) Regional Conference Proceedings*: 171-178

Lepore SJ, Allen KA and Evans GW 1993 Social support lowers cardiovascular reactivity to an acute stressor. *Psychosomatic Medicine 55*: 518-524
http://www.psychosomaticmedicine.org/cgi/reprint/55/6/518?ijkey=e268cc2c49494c9d8eff12dac0b850c3f43e4cf1

Levison PK, Fester CB, Nieman WH and Findley JD 1964 A method for training unrestrained primates to receive drug injection. *Journal of the Experimental Analysis of Behavior 7*: 253-254

Lilly AA, Mehlman PT and Higley J 1999 Trait-like immunological and hematological measures in female rhesus across varied environmental conditions. *American Journal of Primatology 48*: 197-223

Lindburg DG 1971 The rhesus monkey in North India: an ecological and behavioral study. In: Rosenblum LA (ed) *Primate Behavior: Developments in Field and Laboratory Research*, Volume 2 pp. 1-106. Academic Press: New York, NY

Line SW 1987 Environmental enrichment for laboratory primates. *Journal of the American Veterinary Medical Association 190*: 854-859

Line SW, Clarke AS and Markowitz H 1987 Plasma cortisol of female rhesus monkeys in response to acute restraint. *Laboratory Primate Newsletter 26*(4): 1-3
http://www.brown.edu/Research/Primate/lpn26-4.html#line

Line SW, Markowitz H, Morgan KN and Strong S 1989 Evaluation of attempts to enrich the environment of single-caged non-human primates. In: Driscoll JW (ed) *Animal Care and Use in Behavioral Research: Regulation, Issues, and Applications* pp. 103-117. Animal Welfare Information Center, National Agricultural Library: Beltsville, MD

Line SW, Morgan KN, Markowitz H, Roberts J and Riddell M 1990a Behavioral responses of female long-tailed macaques (*Macaca fascicularis*) to pair formation. *Laboratory Primate Newsletter 29*(4): 1-5
http://www.brown.edu/Research/Primate/lpn29-4.html#line

Line SW, Morgan KN, Roberts JA and Markowitz H 1990b Preliminary comments on resocialization of aged macaques. *Laboratory Primate Newsletter 29*(1): 8-12
http://www.brown.edu/Research/Primate/lpn29-1.html#line

Line SW, Markowitz H, Morgan KN and Strong S 1991 Effect of cage size and environmental enrichment on behavioral and physiological responses of rhesus macaques to the stress of daily events. In: Novak MA and Petto AJ (eds) *Through the Looking Glass. Issues of Psychological Well-being in Captive Nonhuman Primates* pp. 160-179. American Psychological Association: Washington DC

Luck CP and Keeble SA 1967 African Monkeys. In: UFAW [Universities Federation for Animal Welfare] (ed) *The UFAW Handbook on the Care and Management of Laboratory Animals* (Third Edition) pp. 734-742. Churchill Livingstone: London, UK

Lueders B 2004 Inside the monkey house. *Isthmus 29*(23): 13-16

Lutz CK and Farrow RA 1996 Foraging device for singly housed longtailed macaques does not reduce stereotypies. *Contemporary Topics in Laboratory Animal Science 35*(3): 75-78

Lutz CK, Chase WK and Novak MA 2000a Abnormal behavior in singly-housed *Macaca mulatta:* Prevalence and potential risk factors. *American Journal of Primatology 51*(Supplement): 71

Lutz C, Tiefenbacher S, Jorgenson MJ, Meyer JS and Novak MA 2000b Techniques for collecting saliva from awake, unrestrained, adult monkeys for cortisol assay. *American Journal of Primatology 52*: 93-99

Lutz CK, Tiefenbacher S, Pouliot AL, Kaufman BM, Meyer JS and Novak MA 2002 Spontaneous episodes of self-biting and associated cortisol levels in captive *Macaca mulatta*. *American Journal of Primatology 57*(Supplement): 43

Lynch JJ and Gantt WH 1968 The heart rate component of the social reflex in dogs: the conditional effects of petting and person. *Conditional Reflex 3*: 69-80

Lynch JJ, Fregin GF, Mackie JB and Monroe RR 1974 Heart rate changes in the horse to human contact. *Psychophysiology 11*: 472-478

Lynch JJ, Thomas SA, Paskewitz DA, Katchar AH and Weir LO 1977 Human contact and cardiac arrhythmia in a coronary care unit. *Psychosomatic Medicine 39*: 188-192
http://www.psychosomaticmedicine.org/cgi/reprint/39/3/188

Lynch JJ 1987 W. Horsley Gantt's Effect of Person. In: McGuigan FJ and Ban TA (eds) *Critical Issues in Psychology, Psychiatry and Physiology* 93. Gordon & Breach: New York, NY

Lynch R 1998 Successful pair-housing of male macaques (*Macaca fascicularis*). *Laboratory Primate Newsletter 37*(1): 4-5
http://www.brown.edu/Research/Primate/lpn37-1.html#pair

Malinow MR, Hill JD and Ochsner AJ 1974 Heart rate in caged rhesus monkeys *(Macaca mulatta)*. *Laboratory Animal Science 24*: 537-540

Manning PJ, Lehner NDM, Feldner MA and Bullock BC 1969 Selected hematologic, serum chemical, and arterial blood gas characteristics of squirrel monkeys *(Saimiri sciureus)*. *Laboratory Animal Care [Laboratory Animal Science] 19*: 831-837

Manuck SB, Kaplan JR and Clarkson TB 1983 Behavioral induced heart rate reactivity and atherosclerosis in cynomolgus monkeys. *Psychosomatic Medicine 45*: 95-108

Manuck SB, Kaplan JR and Matthews KA 1986 Behavioral antecedents of coronary heart disease and atherosclerosis. *Arteriosclerosis 6*: 2-14

Maple T, Erwin J and Mitchell G 1973 Short- and long-term attachments in adult heterosexual pairs of rhesus monkeys. *Paper presented at the Annual Meeting of the Western Psychological Association held in Anaheim, CA*

Markowitz H 1979 Environmental enrichment and behavioral engineering for captive primates. In: Erwin J, Maple T and Mitchell G (eds) *Captivity and Behavior* pp. 217-238. Van Nostrand Reinhold: New York, NY

Marks D, Kelly J, Rice T, Ames S, Marr R, Westfall J, Lloyd J and Torres C 2000 Utilizing restraint chair training to prepare primates for social housing. *Laboratory Primate Newsletter 39*(4): 9-10
http://www.brown.edu/Research/Primate/lpn39-4.html#chair

Marshall K, Jenne M, Maginnis G and Coleman K 2002 Novel environmental enrichment program at the Oregon National Primate Research Center. *Contemporary Topics in Laboratory Animal Science 41*: 98

Mason WA 1960 Socially mediated reduction in emotional responses of young rhesus monkeys. *Journal of Abnormal and Social Psychology 60*: 100-110

McCully CL and Godwin KS 1992 The collar and snaphook restraint system for rhesus monkeys: A new approach to pole and collar training and access port presentation. *Contemporary Topics in Laboratory Animal Science 31*(5): 14-16

McGinnis PR and Kraemer HC 1979 The Stanford outdoor primate facility. *Comfortable Quarters for Laboratory Animals* (Seventh Edition) pp. 20-27. Animal Welfare Institute: Washington, DC

McKinley J, Buchanan-Smith HM, Bassett L and Morris K 2003 Training common marmosets (*Callithrix jacchus*) to cooperate during routine laboratory procedures: Ease of training and time investment. *Journal of Applied Animal Welfare Science 6*: 209-220

McMillan JL, Maier A and Coleman K 2004 Pair housing adult female rhesus macaques: Is it always the best option? *Folia Primatologica 75*(Supplement): 395

McNulty JA, Iskandar E and Kyes RC 2004 Time budgets of long-tailed macaques (*Macaca fascicularis*) on Tinjil Island, Indonesia. *American Journal of Primatology 62* (Supplement): 65-66
http://www.asp.org/asp2004/abstractDisplay.cfm?abstractID=715&confEventID=765

Medical Research Council 2004 *MRC Ethics Guide: Best Practice in the Accomodation and Care of Primates used in Scientific Research*. Medical Research Council: London, UK
http://www.mrc.ac.uk/pdf-primate-best-practice.pdf

Melhuish S and Payne H 2006 Nurses' attitudes to pain management during routine venipuncture in young children. *Paediatric Nursing 18*(2): 20-23

Menzel EW 1991 Chimpanzees *(Pan troglodytes):* Problem seeking versus the bird-in-hand, least-effort strategy. *Primates 32*: 497-508

Michael RP, Setchell KDR and Plant TM 1974 Diurnal changes in plasma testosterone and studies on plasma corticosteroids in non-anaesthetized male rhesus monkeys (*Macaca mulatta*). *Journal of Endocrinology 63*: 325-335

Minkel R 2007 Pair-housing eliminates compulsive hair-pulling: a case report. *Laboratory Animal Refinement & Enrichment Forum (electronic discussion group)*: September 27, 2007 http://groups.yahoo.com/group/LAREF/members

Mitchell DS, Wigodsky HS, Peel HH and McCaffrey TA 1980 Operant conditioning permits voluntary, noninvasive measurement of blood pressure in conscious, unrestrained baboons *(Papio cynocephalus)*. *Behavior Research Methods and Instrumentation 12*: 4l92-498

Mitchell G and Gomber J 1976 Moving laboratory rhesus monkeys *(Macaca mulatta)* to unfamiliar home cages. *Primates 17*: 543-547

Morgan KN, Line SW and Markowitz H 1998 Zoos, enrichment, and the skeptical observer. In: Shepherdson DH, Mellen JD and Hutchins M (eds) *Second Nature - Environmental Enrichment for Captive Animals* pp. 153-171. Smithsonian Institution Press: Washington, DC

Murchison MA 1995 Forage feeder box for single animal cages. *Laboratory Primate Newsletter 34*(1): 1-2
http://www.brown.edu/Research/Primate/lpn34-1.html#forage

Murray L, Hartner M and Clark LP 2002 Enhancing postsurgical recovery of pair-housed nonhuman primates (*M. fascicularis*). *Contemporary Topics in Laboratory Animal Science 41*(4): 112-113

Nahon NS 1968 A device and techniques for the atraumatic handling of the sub-human primate. *Laboratory Animal Care [Laboratory Animal Science] 18*: 486-487

Nakamichi M and Asanuma K 1998 Behavioral effects of perches on group-housed adult female Japanese monkeys. *Perceptual and Motor Skills 87*: 707-714

Nakamichi M and Yamada K 2007 Long-term grooming partnerships between unrelated adult females in a free-ranging group of Japanese monkeys (*Macaca fuscata*). *American Journal of Primatology 69*: 652-663

National Health and Medical Research Council - Animal Welfare Committee 1997 *Policy on the Use of Non-Human Primates in Medical Research*. National Health and Medical Research Council: Canberra, Australia
http://www7.health.gov.au/nhmrc/ethics/animal/issues/nonhuman.htm

National Research Council 1985 *Guide for the Care and Use of Laboratory Animals* (Sixth Edition). National Institutes of Health: Bethesda, MD

National Research Council 1996 *Guide for the Care and Use of Laboratory Animals* (Seventh Edition). National Academy Press: Washington, DC
http://www.nap.edu/books/0309053773/html/

National Research Council 1998 *The Psychological Well-Being of Nonhuman Primates*. National Academy Press: Washington, DC
http://books.nap.edu/books/0309052335/html/index.html

National Research Council 2008 *Recognition and Alleviation of Distress in Laboratory Animals* (Prepublication Copy). National Academy Press: Washington, DC
http://books.nap.edu/openbook.php?record_id=11931

Nelms R, Davis BK, Tansey G and Raber JM 2001 Utilization of training techniques to minimize distress and facilitate the treatment of a chronically ill macaque. *AALAS [American Association for Laboratory Animal Science] 52nd National Meeting Official Program*: 97-98

Nerem RM, Levensque MJ and Cornhill JF 1980 Social environment as a factor of diet induced atherosclerosis. *Science 208*: 1475-1476

Neveu H and Deputte BL 1996 Influence of availability of perches on the behavioral well-being of captive, group-living mangabeys. *American Journal of Primatology 38*: 175-185

Niemeyer C, Gray EG and Stephen T 1996 Improving the psychological well-being of nonhuman primates by providing appropriate therapeutic devices *XVIth Congress of the International Primatological Society/XIXth Conference of the American Society of Primatologists, Abstract No. 678*

Nishida T and Wallauer W 2003 Leaf-pile pulling: An unusual play pattern in wild chimpanzees. *American Journal of Primatology 60*: 167-173

Novak MA 2003 Self-injurious behavior in rhesus monkeys: New insights into its etiology, physiology, and treatment. *American Journal of Primatology 59*: 3-19

Novak MA and Suomi SJ 1988 Psychological well-being of primates in captivity. *American Psychologist 43*: 765-773

O'Connor E and Reinhardt V 1994 Caged stumptailed macaques voluntarily work for ordinary food. *In Touch* 1(1): 10-11
http://www.awionline.org/Lab_animals/biblio/tou-food.htm

Office of Laboratory Animal Welfare (OLAW) 2002 *Institutional Animal Care and Use Committee Guidebook* (Second Edition). Office of Laboratory Animal Welfare: Bethesda, MD

Paquette D 1992 Discovering and learning tool-use for fishing honey by captive chimpanzees. *Human Evolution* 7(3): 17-30

Panneton M, Alleyn S and Kelly N 2001 Chair restraint for squirrel monkeys. *AALAS [American Association for Laboratory Animal Science] 52nd National Meeting Official Program*: 92

Pastorello L 1998 Enrichment of a blind monkey. In: Hare VJ and Worley E (eds) *Proceedings of the Third International Conference on Environmental Enrichment* pp. 374-379. The Shape of Enrichment: San Diego, CA

Pedersen V, Barnett JL, Hemsworth PH, Newman EA and Schirmer B 1998 The effects of handling on behavioural and physiological responses to housing in tether-stalls among pregnant pigs. *Animal Welfare* 7: 137-150

Phoenix CH and Chambers KC 1984 Sexual behavior and serum hormone levels in aging rhesus males: Effects of environmental change. *Hormones and Behavior 18*: 206-215

Platt DM, Kinsey JH, Jorgenson MJ and Novak MA 1996 Factors affecting the expression of self-injurious behavior in rhesus monkeys *(Macaca mulatta). XVIth Congress of the International Primatological Society/XIXth Conference of the American Society of Primatologists, Abstract No. 768*

Poole T, Hubrecht R and Kirkwood JK 1999 Marmosets and Tamarins. In: UFAW [Universities Federation for Animal Welfare] (edited by Poole, T. and English, P.) *The UFAW Handbook on the Care and Management of Laboratory Animals* (Seventh Edition) pp. 558-573. Blackwell Science: Oxford, UK

Pranger LA, Maier A, Coleman K, Lambeth SP, Perlman JE, Thiele E, McMillam JL and Schapiro SJ 2006 Venipuncture training using positive reinforcement training techniques: a comparison of chimpanzee and rhesus macaques. *American Journal of Primatology 68* (Supplement): 61-62

Prentice ED, Zucker IH and Jameton A 1986 Ethics of animal welfare in research: The institution's attempt to achieve appropriate social balance. *The Physiologist 29*: 1&19-21

Prescott MJ 2002 Ideal goals for training of laboratory personnel to work with primates. *XIXth Congress of the International Primatological Society, Abstracts*: 185

Priest GM 1990 The use of operant conditioning in training husbandry behavior with captive exotic animals. *Proceedings of the National American Association of Zoo Keepers Conference 16*: 94-108

Priest GM 1991a Training a diabetic drill (*Mandrillus leucophaeus*) to accept insulin injections and venipuncture. *Laboratory Primate Newsletter 30*(1): 1-4
http://www.brown.edu/Research/Primate/lpn30-1.html#loon

Priest, GM 1991b *Loon, the Diabetic Drill* (Videotape with commentary). Mac & Mutley Kpix (CBS): San Francisco, CA

Primate Research Institute 2003 *Guide of the Care and Use of Laboratory Primates* (Second Edition). Primate Research Institute of Kyoto University: Kyoto, Japan
http://www.pri.kyoto-u.ac.jp/research/en/

Quadri SK, Pierson C and Spies HP 1978 Effects of centrally acting drugs on serum levels in rhesus monkeys. *Neuroendocrinology 27*: 136-147

Ranheim S and Reinhardt V 1989 Compatible rhesus monkeys provide long-term stimulation for each other. *Laboratory Primate Newsletter 28*(3): 1-2
http://www.brown.edu/Research/Primate/lpn28-3.html#vik

Rasmussen KLR 1985 Response to social separation in adult macaques. *American Journal of Primatology 8*: 358-359

Ratajeski MA and McDonald KM 2005 The use of polycarbonate privacy panels to reduce stress in singly housed macaques. *AALAS [American Association for Laboratory Animal Science] 56th National Meeting Official Program*: 135

Reichard T and Shellaberger W 1992 Training for husbandry and medical procedures. *American Zoo and Aquarium Association (AZA) Annual Conference Proceedings*: 396-402

Reinhardt V 1989a Behavioral responses of unrelated adult male rhesus monkeys familiarized and paired for the purpose of environmental enrichment. *American Journal of Primatology 17*: 243-248
http://www.brown.edu/Research/Primate/lpn27-4.html#vik

Reinhardt V 1989b Evaluation of the long-term effectiveness of two environmental enrichment objects for singly caged rhesus macaques. *Lab Animal 18*(6): 31-33
http://www.awionline.org/Lab_animals/biblio/la-eval.htm

Reinhardt V 1990a Social enrichment for laboratory primates: A critical review. *Laboratory Primate Newsletter 29*(3): 7-11
http://www.brown.edu/Research/Primate/lpn29-3.html#rev

REFERENCES

Reinhardt V 1990b Time budget of caged rhesus monkeys exposed to a companion, a PVC perch and a piece of wood for an extended time. *American Journal of Primatology 20*: 51-56

Reinhardt V 1990c Comparing the effectiveness of PVC perches versus wooden perches as environmental enrichment objects for singly caged rhesus monkeys. *Laboratory Primate Newsletter 29*(1): 13-14
http://www.brown.edu/Research/Primate/lpn29-1.html#comp

Reinhardt V 1991a Social enrichment for aged rhesus monkeys who have lived singly for many years. *Animal Technology 43*: 173-177
http://www.awionline.org/Lab_animals/biblio/at173.htm

Reinhardt V 1991b Group formation of previously single-caged adult rhesus macaques for the purpose of environmental enrichment. *Journal of Experimental Animal Science 34*: 110-115
http://www.awionline.org/Lab_animals/biblio/es34-1.html

Reinhardt V 1991c An environmental enrichment program for caged rhesus monkeys at the Wisconsin Regional Primate Research Center. In: Novak MA and Petto AJ (eds) *Through the Looking Glass. Issues of Psychological Well-being in Captive Nonhuman Primates* pp. 149-159. American Psychological Association: Washington, DC

Reinhardt V 1991d Training adult male rhesus monkeys to actively cooperate during in-homecage venipuncture. *Animal Technology 42*: 11-17
http://www.awionline.org/Lab_animals/biblio/at11.htm

Reinhardt V 1992a Avoiding aggression during and after pair formation of adult rhesus macaques. *Laboratory Primate Newsletter 31*(3): 10-11
http://www.brown.edu/Research/Primate/lpn31-3.html#avoid

Reinhardt V 1992b Difficulty in training juvenile rhesus macaques to actively cooperate during venipuncture in the homecage. *Laboratory Primate Newsletter 31*(3): 1-2
http://www.brown.edu/Research/Primate/lpn31-3.html#diff

Reinhardt V 1992c Transport-cage training of caged rhesus macaques. *Animal Technology 43*: 57-61
http://www.awionline.org/Lab_animals/biblio/at57.htm

Reinhardt V 1993a Enticing nonhuman primates to forage for their standard biscuit ration. *Zoo Biology 12*: 307-312
http://www.awionline.org/Lab_animals/biblio/zb12-30.htm

Reinhardt V 1993b Evaluation of an inexpensive custom-made food puzzle used as primary feeder for pair-housed rhesus macaques. *Laboratory Primate Newsletter 32*(3): 7-8
http://www.brown.edu/Research/Primate/lpn32-3.html#food

Reinhardt V 1993c Promoting increased foraging behaviour in caged stumptailed macaques. *Folia Primatologica 61*: 47-51

Reinhardt V 1993d Using the mesh ceiling as a food puzzle to encourage foraging behaviour in caged rhesus macaques (*Macaca mulatta*). *Animal Welfare 2*: 165-172
http://www.awionline.org/Lab_animals/biblio/aw3mesh.htm

Reinhardt V 1994a Caged rhesus macaques voluntarily work for ordinary food. *Primates 35*: 95-98
http://www.awionline.org/Lab_animals/biblio/primat~1.htm

Reinhardt V 1994b Pair-housing rather than single-housing for laboratory rhesus macaques. *Journal of Medical Primatology 23*: 426-431
http://www.awionline.org/Lab_animals/biblio/jmp23.htm

Reinhardt V 1994c Social enrichment for previously single-caged stumptail macaques. *Animal Technology 5*: 37-41
http://www.awionline.org/Lab_animals/biblio/at37.htm

Reinhardt V 1997 Refining the traditional housing and handling of laboratory rhesus macaques improves scientific methodology. *Primate Report 49*: 93-112
http://www.awionline.org/Lab_animals/biblio/pr-refi.htm

Reinhardt V 1998 Housing and handling of nonhuman primates. In: Bekoff M and Meaney C (eds) *Encyclopedia of Animal Rights and Animal Welfare* pp. 217-222. Greenwood Press: Westport, CT

Reinhardt V 1999 Pair-housing overcomes self-biting behavior in macaques. *Laboratory Primate Newsletter 38*(1): 4
http://www.brown.edu/Research/Primate/lpn38-1.html#pair

Reinhardt V 2003a Legal loophole for subminimal floor area for caged macaques. *Journal of Applied Animal Welfare Science 6*: 53-56
http://www.awionline.org/Lab_animals/biblio/jaaws9.html

Reinhardt V 2003b Working with rather than against macaques during blood collection. *Journal of Applied Animal Welfare Science 6*: 189-197
http://www.awionline.org/Lab_animals/biblio/jaaws11.html

Reinhardt, V and Dodsworth, R 1989 *Facilitated Socialization of Previously Single-Caged Adult Rhesus Macaques* (Videotape with commentary). Wisconsin Regional Primate Research Center: Madison, WI
http://www.awionline.org/Lab_animals/biblio/vid-vik.htm

Reinhardt V and Cowley D 1990 Training stumptailed monkeys to cooperate during in-homecage treatment. *Laboratory Primate Newsletter 29*(4): 9-10
http://www.brown.edu/Research/Primate/lpn29-4.html#vik

Reinhardt V and Pape R 1991 An alternative method for primate perch installation. *Lab Animal 20*(8): 47-48
http://www.awionline.org/Lab_animals/biblio/la-an.htm

Reinhardt V and Reinhardt A 1991 Impact of a privacy panel on the behavior of caged female rhesus monkeys living in pairs. *Journal of Experimental Animal Science 34*: 55-58
http://www.awionline.org/Lab_animals/biblio/es34-5~1.htm

Reinhardt V and Cowley D 1992 In-homecage blood collection from conscious stumptailed macaques. *Animal Welfare 1*: 249-255
http://www.awionline.org/Lab_animals/biblio/aw1blood.htm

Reinhardt V and Hurwitz S 1993 Evaluation of social enrichment for aged rhesus macaques. *Animal Technology 44*: 53-57
http://www.awionline.org/Lab_animals/biblio/at44.htm

Reinhardt, V and Reinhardt, A 2001 *Environmental Enrichment for Caged Rhesus Macaques (Macaca mulatta) - Photographic Documentation and Literature Review* (Second Edition). Animal Welfare Institute: Washington, DC
http://www.awionline.org/lab_animals/rhesus/Photo.htm

Reinhardt V, Houser WD, Eisele S and Champoux M 1987 Social enrichment with infants of the environment for singly caged adult rhesus monkeys. *Zoo Biology 6*: 365-371
http://www.awionline.org/Lab_animals/biblio/zb6-365.htm

Reinhardt V, Houser WD, Eisele S, Cowley D and Vertein R 1988a Behavior responses of unrelated rhesus monkey females paired for the purpose of environmental enrichment. *American Journal of Primatology 14*: 135-140
http://www.brown.edu/Research/Primate/lpn26-2.html#vik

Reinhardt V, Cowley D, Eisele S, Vertein R and Houser WD 1988b Pairing compatible female rhesus monkeys for the purpose of cage enrichment has no negative impact on body weight. *Laboratory Primate Newsletter 27*(1): 13-15
http://www.brown.edu/Research/Primate/lpn27-1.html#pair

Reinhardt V, Houser WD and Eisele S 1989 Pairing previously singly caged rhesus monkeys does not interfere with common research protocols. *Laboratory Animal Science 39*: 73-74

Reinhardt V, Cowley D, Scheffler J and Vertein R 1990a Living continuously with a compatible companion is not a distressing experience for rhesus monkeys. *Laboratory Primate Newsletter 29*(2): 16-17
http://www.brown.edu/Research/Primate/lpn29-2.html#rhesus

Reinhardt V, Cowley D, Scheffler J, Vertein R and Wegner F 1990b Cortisol response of female rhesus monkeys to venipuncture in homecage versus venipuncture in restraint apparatus. *Journal of Medical Primatology 19*: 601-606
http://www.awionline.org/Lab_animals/biblio/jmp19.htm

Reinhardt V, Liss C and Stevens C 1995 Restraint methods of laboratory nonhuman primates: A critical review. *Animal Welfare 4*: 221-238
http://www.awionline.org/Lab_animals/biblio/aw6metho.htm

Richmond CA, Ross NA and Egeland GM 2007 Social support and thriving health: a new approach to understanding the health of indigenous Canadians. *American Journal of Public Health 97*: 1827-1833

Robbins DQ, Zwick H, Leedy M and Stearns G 1986 Acute restraint device for rhesus monkeys. *Laboratory Animal Science 36*: 68-70

Roberts SJ and Platt ML 2005 Effects of isosexual pair-housing on biomedical implants and study participation in male macaques. *Contemporary Topics in Laboratory Animal Science 44*(5): 13-18
http://www.aalas.org/pdfUtility.aspx?pdf=CT/44_05_02.pdf

Rosenberg DP and Kesel ML 1994 Old-World monkeys. In: Rollin BE and Kesel ML (eds) *The Experimental Animal in Biomedical Research: Care, Husbandry, and Well-Being - An Overview by Species* pp. 457-483. CPR Press: Boca Raton, FL

Rosenblum IY and Coulston F 1981 Normal range of longitudinal blood chemistry and hematology values in juvenile and adult rhesus monkeys (*Macaca mulatta*). *Ecotoxicology and Environmental Safety 5*: 401-411

Rowell TE and Hinde RA 1963 Responses of rhesus monkeys to mildly stressful situations. *Animal Behaviour 11*: 235-243

Russell C and Russell WMS 1985 Conflict activities in monkeys. *Social Biology and Human Affairs 50*: 26-48

Russell WMS 2005 The three Rs: past, present and future. *Animal Welfare 14*: 279-286

Russell WMS and Burch RL 1992 *The Principles of Humane Experimental Technique* (Special Edition). Wheathampstead, UK: Universities Federation for Animal Welfare - Originally published in 1959 by Methuen & Co. London, UK
http://altweb.jhsph.edu/publications/humane_exp/het-toc.htm

Sainsbury AW, Mew JA, Purton P, Eaton BD and Cooper JE 1990 Advances in the management of primates kept for biomedical research. *Animal Technology 41*: 87-101

Sauceda R and Schmidt MG 2000 Refining macaque handling and restraint. *Lab Animal 29*(1): 47-49

Scallet AC, McKay D, Bailey JR, Ali SF, Paule MG, Slikker W and Rayford PL 1989 Meal-induced increase in plasma gastrin immunoreactivity in the rhesus monkey (*Macaca mulatta*). *American Journal of Primatology 18*: 315-319

Schapiro SJ 2005 Chimpanzees used in research: Voluntary blood samples differ from anesthetized samples. *AWI Quarterly 54*(3): 15-16
http://www.awionline.org/pubs/Quarterly/05_54_03/05_54_3p15_6.htm

Schapiro SJ and Bushong D 1994 Effects of enrichment on veterinary treatment of laboratory rhesus macaques (*Macaca mulatta*). *Animal Welfare 3*: 25-36
http://www.awionline.org/Lab_animals/biblio/aw3-25.htm

Schapiro SJ, Nehete PN, Perlman JE and Sastry KJ 1997 Social housing condition affects cell-mediated immune responses in adult rhesus macaques. *American Journal of Primatology 42*: 147

Schapiro SJ, Nehete PN, Perlman JE and Sastry KJ 2000 A comparison of cell-mediated immune responses in rhesus macaques housed singly, in pairs, or in groups. *Applied Animal Behaviour Science 68*: 67-84

Schmidt EM, Dold GM and McIntosh JS 1989a A perch for primate squeeze cages. *Laboratory Animal Science 39*: 166-167

Schmidt EM, Dold GM and McIntosh JS 1989b A simple transfer and chairing technique for nonhuman primates. *Laboratory Animal Science 39*: 258-260

Schnell CR 1997 Haemodynamic measurements by telemetry in conscious unrestrained marmosets: Responses to social and non social stress events. In: Pryce C, Scott L and Schnell C (ed) *Marmosets and Tamarins in Biological and Biomedical Research. Proceedings of a Workshop* pp. 170-180. DSSD Imagery: Salisbury, UK

Schnell CR and Wood JM 1993 Measurement of blood pressure, heart rate, body temperature, ECG and activity by telemetry in conscious unrestrained marmosets. *Proceedings of the Fifth Federation of European Laboratory Animal Science Associations (FELASA) Symposium*: 107-111

Scott GD and Gendreau P 1969 Psychiatric implications of sensory deprivation in a maximum security prison. *Canadian Psychiatric Association Journal 14*: 337-341

Scott LAM, Pearce PC, Fairhall S, Muggleton NG and Smith JN 2002 Training non-human primates to cooperate with scientific procedures in applied biomedical research. *XIXth Congress of the International Primatological Society, Abstracts*: 183

Selekman J and Snyder B 1996 Uses of and alternatives to restraints in pediatric settings. *AACN [American Association of Critical-Care Nurses] Clinical Issues 7*: 603-610

Shibata C and Ford SM 2007 The maintenance of social bonds in adult pairs of captive cotton-top tamarins (*Saguinus oedipus*). *American Journal of Physical Anthropology, Supplement* (44): 217

Shear K and Shair H 2005 Attachment, loss, and complicated grief. *Developmental Psychobiology 47*: 253-267

Shively CA, Clarkson TB and Kaplan JR 1989 Social deprivation and coronary artery atherosclerosis in female cynomolgus monkeys. *Atherosclerosis 77*: 69-76

Shumaker, SA and Czajkowski, SM 1994 *Social Support and Cardiovascular Disease*. Plenum Press: New York, NY

Siegel JM 1990 Stressful life events and use of physician services among the elderly: the moderating role of pet ownership. *Journal of Personality and Social Psychology 58*: 1081-1086

Silk JB 2003 Cooperation without counting: The puzzle of friendship. In: Hammerstein P (ed) *Genetic and Cultural Evolution of Cooperation* pp. 37-54. MIT Press/Dahlem University Press: Cambridge, MA

Silk JB, Alberts SC and Altmann J 2006 Family ties, friendship, and fitness among wild female baboons. *International Journal of Primatology 27* (Supplement): 530

Skoumbourdis EK 2008 Pole-and-collar-and-chair training. *Laboratory Animal Refinement & Enrichment Forum* (electronic discussion group): January 24, 2008 http://groups.yahoo.com/group/LAREF/members

Sluga W GJ 1969 Self-inflicted injury and self-mutilation in prisoners [article in German]. *Wiener Medizinische Wochenschrift 119*: 453-459

Smith K, St. Claire M, Shaver C and Olexa P 2004 Use of a shredded paper substrate to ameliorate abnormal self-directed behavior of a chimpanzee (*Pan troglodytes*). *American Journal of Primatology 62*(Supplement): 94-95 http://www.asp.org/asp2004/abstractDisplay.cfm?abstractID=702&confEventID=825

Smith TE and French JA 1997 Separation-induced activity in the hypothalamic-pituitary adrenal axis in a social primate (*Callithrix kuhli*). *American Journal of Primatology 42*: 150

Smuts B 2004 Friendship in animals. In: Bekoff M (ed) *Encyclopedia of Animal Behavior Volume 2* pp. 599-602. Greenwood Press: Westport, CT

Southwick CH, Beg MA and Siddiqi MR 1965 Rhesus monkeys in North India. In: De Vore I (ed) *Primate Behavior - Field Studies of Monkeys and Apes* pp. 111-159. Holt, Rinehart and Winston: New York, NY

Sotir M, Switzer W, Schable C, Schmitt J, Vitek C and Khabbaz RF 1997 Risk of occupational exposure to potentially infectious nonhuman primate materials and to immunodeficiency virus. *Journal of Medical Primatology 26*: 233-240

Spector M, Kowalczky MA, Fortman JD and Bennett BT 1994 Design and implementation of a primate foraging tray. *Contemporary Topics in Laboratory Animal Science 33*(5): 54-55

Spiegel D and Sephton SE 2001 Psychoneuroimmune and endocrine pathways in cancer: effects of stress and support. *Journal of Clinical Neuropsychiatry*: 252-265

Stopka P, Johnson DDP and Barrett L 2001 'Friendship' for fitness or 'friendship' for friendship's sake? *Animal Behaviour Forum 61*(3): F19-F21 http://www.academicpress.com/www/journal/ar/1600a.pdf

Storey PL, Turner PV and Tremblay JL 2000 Environmental enrichment for rhesus macaques: A cost-effective exercise cage. *Contemporary Topics in Laboratory Animal Science 39*(1): 14-16

Stringfield CE and McNary JK 1998 Operant conditioning for diabetic primates to accept insulin injections. *American Association of Zoo Veterinarians (AAZV)/American Association of Wildlife Veterinarians (AAWV) Joint Conference Proceedings*: 396-397

Strum SC 1985 Baboons may be smarter than people: For these political primates, friendship - not aggression - is the key to survival. *Animal Kingdom 88*(2): 12-25

Suedfeld P 1984 Measuring the effects of solitary confinement. *American Journal of Psychiatry 141*: 1306-1308

Suleman MA, Njugana J and Anderson J 1988 Training of vervet monkeys, sykes monkeys and baboons for collection of biological samples. *Proceedings of the XIIth Congress of the International Primatological Society*: 12

Teas J, Richie T, Taylor H and Southwick C 1980 Population patterns and behavioral ecology of rhesus monkeys *(Macaca mulatta)* in Nepal. In: Lindburgh DG (ed) *The Macaques: Studies in Ecology, Behavior and Evolution* pp. 247-262. Van Nostrand Reinhold: New York, NY

Thorsteinsson EB, James JE and Gregg ME 1998 Effects of video-relayed social support on hemodynamic reactivity and salivary cortisol during laboratory-based behavioral challenge. *Health Psychology 17*: 436-444

Tiefenbacher S, Lee B, Meyer JS and Spealman RD 2003 Noninvasive technique for the repeated sampling of salivary free cortisol in awake, unrestrained squirrel monkeys. *American Journal of Primatology 60*: 69-75

Todd-Schuelke S, Trask B, Wallace C, Baun MM, Bergstrom N and McCabe B 1991/92 Physiological effects of the use of a companion animal dog as a cue to relaxation in diagnosed hypertensives. *Delta Society* (Web site; accessed January 16, 2008) http://www.deltasociety.org/AnimalsHealthAdultsPhysiological.htm

Tomlinson D 2004 Physical restraint during procedures: issues and implications for practice. *Journal of Pediatric Oncology Nursing 21*: 258-263

Tully LA 2003 Paint Roller Enrichment. *Tech Talk [The Newsletter for Laboratory Animal Science Technicians] 8*(3): 1-2

Tully LA, Jenne M and Coleman K 2002 Paint roller and grooming-boards as treatment for over-grooming rhesus macaques. *Contemporary Topics in Laboratory Animal Science 41*(4): 75

Turkkan JS 1990 New methodology for measuring blood pressure in awake baboons with use of behavioral training techniques. *Journal of Medical Primatology 19*: 455-466 http://www.awionline.org/Lab_animals/biblio/jmp19-4.htm

Turkkan JS, Ator NA, Brady JV and Craven KA 1989 Beyond chronic catheterization in laboratory primates. In: Segal EF (ed) *Housing, Care and Psychological Wellbeing of Captive and Laboratory Primates* pp. 305-322. Noyes Publications: Park Ridge, NJ

Turner P and Grantham LE 2002 Short-term effects of an environmental enrichment program for adult cynomolgus monkeys. *Contemporary Topics in Laboratory Animal Science 41*: 13-17

Tustin GW, Williams LE and Brady AG 1996 Rotational use of a recreational cage for the environmental enrichment of Japanese macaques (*Macaca fuscata*). *Laboratory Primate Newsletter 35*(1): 5-7 http://www.brown.edu/Research/Primate/lpn35-1.html#tustin

Uchino BN, Cacioppo JT and Kiecolt-Glaser JK 1996 The relationship between social support and physiological processes: a review with emphasis on underlying mechanisms and implications for health. *Psychological Bulletin 119*: 488-531

US Department of Agriculture 1995 *Code of Federal Regulations, Title 9, Chapter 1, Subchapter A - Animal Welfare*. U.S. Government Printing Office: Washington, DC http://www.aphis.usda.gov/ac/9CFR99.html

REFERENCES

US Department of Agriculture 1999 *Final Report on Environmental Enhancement to Promote the Psychological Well-Being of Nonhuman Primates*. U.S. Department of Agriculture - Animal Care: Riverdale, MD
http://www.aphis.usda.gov/ac/eejuly15.html

Uno D, Uchino BN and Smith TW 2002 Relationship quality moderates the effect of social support given by close friends on cardiovascular reactivity in women. *International Journal of Behavioral Medicine 9*: 243-262

Valerio, DA, Miller, RL, Innes, JRM, Courntey, KD, Pallotta, AJ and Guttmacher, RM 1969 *Macaca mulatta. Management of a Laboratory Breeding Colony*. Academic Press: New York, NY

Van Lawick-Goodall J 1968 A preliminary report on expressive movements and communication in the Gombe Stream chimpanzee. In: Jay PC (ed) *Primates - Studies in Adaptation and Variability* pp. 313-374. Holt, Rinehart and Winston: New York, NY

Villalba R and Harrington C 2003 Repetitive self-injurious behavior: The emerging potential of psychotropic intervention. *Psychiatric Times 20*(2): 1-11
http://www.psychiatrictimes.com/p030266.html

Vogt JL, Coe CL and Levine S 1981 Behavioral and adrenocorticoid responsiveness of squirrel monkeys to a live snake: is flight necessarily stressful? *Behavioral and Neural Biology 32*: 391-405

Vormbrock JK and Grossberg JM 1988 Cardiovascular effects of human-pet dog interactions. *Journal of Behavioral Medicine 11*: 509-517

Wall HS, Worthman C and Else JG 1985 Effects of ketamine anaesthesia, stress and repeated bleeding on the haematology of vervet monkeys. *Laboratory Animals 19*: 138-144

Walters RH, Callagan JE and Newman AF 1963 Effect of solitary confinement on prisoners. *American Journal of Psychiatry 120*: 771-773

Washburn DA and Rumbaugh DM 1992 Investigations of rhesus monkey video-task performance: Evidence for enrichment. *Contemporary Topics in Laboratory Animal Science 31*(5): 6-10

Watts DP 2007 Long-term stability of male chimpanzee social relationships at Ngogo. *American Journal of Physical Anthropology, Supplement* (44): 245

Watson LM 1992 Effect of an enrichment device on stereotypic and self-aggressive behaviors in singly-caged macaques: A pilot study. *Laboratory Primate Newsletter 31*(3): 8-10
http://www.brown.edu/Research/Primate/lpn31-3.html#watson

Watson LM 2002 A successful program for same- and cross-age pair-housing adult and sub-adult male *Macaca fascicularis. Laboratory Primate Newsletter 41*(2): 6-9 http://www.brown.edu/Research/Primate/lpn41-2.html#watson

Watson LM, Cosby R and Lee-Parritz DE 1993 Behavioral effects of enrichment devices on laboratory primates with stereotypic and self-directed behavior. *American Journal of Primatology 31*: 355-356

Watson SL, Shively CA, Kaplan JR and Line SW 1998 Effects of chronic social separation on cardiovascular disease risk factors in female cynomolgus monkeys. *Atherosclerosis 137*: 259-266

Watson SL, Shively CA and Voytko ML 1999 Can puzzle feeders be used as cognitive screening instruments? Differential performance of young and aged female monkeys on a puzzle feeder task. *American Journal of Primatology 49*: 195-202

Weed JL, Wagner PO, Byrum R, Parrish S, Knezevich M and Powell DA 2003 Treatment of persistent self-injurious behavior in rhesus monkeys through socialization: A preliminary report. *Contemporary Topics in Laboratory Animal Science 42*(5): 21-23

Wells DA 1972 The use of seclusion on a university hospital psychiatric floor. *Archives of General Psychiatry 26*: 410-413

Wheatley, BP 1999 *The Sacred Monkeys of Bali*. Waveland Press: Prospect Heights, IL

Wheeler MD, Schutzengel RE, Barry S and Styne DM 1990 Changes in basal and stimulated growth hormone secretion in the aging rhesus monkeys: A comparison of chair restraint and tether and vest sampling. *Journal of Clinical Endocrinology and Metabolism 71*: 1501-1507

White G, Hill W, Speigel G, Valentine B, Weigant J and Wallis J 2000 Conversion of canine runs to group social housing for juvenile baboons. *AALAS [American Association for Laboratory Animal Science] 51st National Meeting Official Program*: 126

Whitney RA and Wickings EJ 1987 Macaques and other old world simians. In: Poole TB (ed) *The UFAW Handbook on the Care and Management of Laboratory Animals,* (Sixth Edition) pp. 599-627. Churchill Livingstone: New York, NY

Whitney RA, Johnson DJ and Cole WC 1973 *Laboratory Primate Handbook*. Academic Press: New York, NY

Wickings EJ and Nieschlag E 1980 Pituitary response to LRH and TRH stimulation and peripheral steroid hormones in conscious and anaesthetized adult male rhesus monkeys *(Macaca mulatta). Acta Endocrinologica 93*: 287-293

Wilson CC 1987 Physiological responses of college students to a pet. *Journal of Nervous and Mental Disease 175*: 606–612

Witcher SJ and Fisher JD 1979 Multidimensional reaction to therepeutic touch in a hospital setting. *Journal of Personality and Social Psychology 37*: 87-96

Wolfle TL 1987 Control of stress using non-drug approaches. *Journal of the American Veterinary Medical Association 191*: 1219-1221

Woodbeck T and Reinhardt V 1991 Perch use by *Macaca mulatta* in relation to cage location. *Laboratory Primate Newsletter 30*(4): 11-12
http://www.brown.edu/Research/Primate/lpn30-4.html#perch

Wrangham RW 1992 Living naturally: Aspects of wild environments relevant to captive chimpanzee management. In: Erwin J and Landon JC (eds) *Chimpanzee Conservation and Public Health: Environments for the Future* pp. 71-81. Diagnon/Bioqual: Rockville, MD

Wrightsman LS 1960 Effects of waiting with others on changes in level of felt anxiety. *Journal of Abnormal and Social Psychology 61*: 216-222

Yaroshevsky F 1975 Self-mutilation in Soviet prisons. *Canadian Psychiatric Association Journal 20*: 443-446

Zakaria M, Lerche NW, Chomel BB and Kass PH 1996 Accidental injuries associated with nonhuman primate exposure at two Regional Primate Research Centers (U.S.A.): 1988-1993. *Laboratory Animal Science 46*: 298-304